Genes, Genomes and Society

Röbbe Wünschiers

Genes, Genomes and Society

Röbbe Wünschiers
Fakultät für Angewandte
Computer- und Biowissenschaften
Hochschule Mittweida
Mittweida, Germany

The translation was done with the help of artificial intelligence (machine translation by the service DeepL.com). A subsequent human revision was done primarily in terms of content.

ISBN 978-3-662-64080-7   ISBN 978-3-662-64081-4   (eBook)
https://doi.org/10.1007/978-3-662-64081-4

This Springer imprint is published by the registered company Springer-Verlag GmbH, DE part of Springer Nature.
The registered company address is: Heidelberger Platz 3, 14197 Berlin, Germany

Preface

The genetic material of a wide variety of organisms has already been used in many different ways. What does this do to us as a society and what can still be expected? With this book, I would like to venture a retrospective, an overview, and an outlook. So, I always try to span temporal arcs. Since I am addressing interested readers—whom I trust to be quite capable—I not only introduce them to relevant basics, but also try my hand at painting pictures: The technically advanced and critical experts may forgive me for some abstract abstraction. If you thumb through the book, you will encounter not only, but also, stuffed, scientific looking illustrations. This should not deter you—I will try to guide your eyes and thoughts. My aim is to provide a diverse readership with the tools to form their own opinions on the subject of genetics and genomics and its application, genetic engineering. You should be able to position yourself as to how you personally want to deal with the **genetic engineering revolution** and what you expect from representatives from politics, science, and society. **Gene editing** methods, with which we can modify the genetic material of all living beings as never before, would be described by economists as a so-called **disruptive technology.** This means that it displaces existing methods. I therefore believe that it is legitimate to speak of a new generation of researchers, patients, beneficiaries, supporters, and opponents: the **generation of gene editors**. But do not expect a book that is primarily about gene editing. No, it is primarily about the genetic engineering revolution.

Now that you have this book in front of you, I would like to make two wishes. Genetic engineering and its application are the tip of an iceberg called science. It is the result of contributions from many disciplines, what we scientists call interdisciplinary. So, the subject is complex. So, my **first wish** is: please take your time reading, and use encyclopaedias, Wikipedia, uncle Google, or aunt Yahoo when you encounter thought barriers. Since explaining unfamiliar terms has become so easy with the internet, I have not included a glossary. Also talk to friends, ask me, or discuss on *generation-genschere.de*. My **second wish**: Think colourfully and not in black-and-white categories. Sure, sugar makes caries and fat, but it also tastes great and preserves fruits.

One more thing: Do not be scared by **bold print**; it is just to help you find people and key words again. And if you find the **illustrations** too small, check them out in the eBook at screen size.

You read it over and over again and think, sure: now for the **thanks to the** dear partner who put up with the scribbler for so long. But there is nothing to discuss: You cannot do it without time off with little distractions, good food, literature, and joint discussions. Therefore, first and foremost, I thank my wife Catherine from the bottom of my heart. I would also like to thank my colleagues who have supported me as much as possible—first and foremost Sandra Feik, René Kretschmer, Nadine Wappler, Robert Leidenfrost, and Jacqueline Günther. I would like to thank

Josi Hesse for her insight into *fitness, food, and genes* and the staff of the university library in Mittweida for feeding my mind. I thank Sarah Koch at the publishing house for well-dosed, inspiring conversations and creatively constructive e-mail chats. For proofreading and patient editing of comma and point mutations in the German edition, I thank Cornelia Reichert. Finally, it may sound strange, I thank the European Commission for recently rejecting my research proposals and thus giving me the totally unexpected opportunity to deal more intensively with the big picture.

The following **epilogue** from the 1999 book *Biology of the Prokaryotes* should conclude my preface:

> Especially in the beginning, revolutionary new technologies normally lend themselves to controversial discussions. Some people are afraid that these technologies may constitute uncontrollable dangers or may threaten traditional values and techniques. Others argue that because these methods are revolutionary and new, they are extremely promising and hence must be given any opportunity to be developed. Most geneticists and biologists who use recombinant DNA techniques constitute a third, more neutral group for which gene technology is but a logical continuation of the previous developments in genetics as initiated by scientists such as C. Darwin, G. Mendel, and B. McClintock. They are convinced that the biological risks outlined above are not radically new and hence can be handled by using reasonable precautions. In their view, gene technology has shown its outstanding value for basic research in an amazingly short time and will also prove its practical value within reasonable expectations.
>
> Finally, they are aware that gene technology will provoke essential ethical, legal, economic, and social questions due to its potential effects on living organisms, including man.

At a closer look, however, most if not all of these will be recognised as millennium-old questions, which probably still will be asked in millenniums because perhaps they never can be answered definitively and will have to be asked by each new generation as long as there are human beings [1].

And now, dear friends, I look forward to a cup of tea with you.

July 2019   Röbbe Wünschiers
University of Applied Sciences Mittweida
Mittweida, Germany

Reference

1. Lengeler JW, Drews G, Schlegel HG (1999) Biology of the Prokaryotes. Georg Thieme Verlag, Stuttgart. https://doi.org/10.1002/9781444313314

Preface to This English Translation

The German edition of this book was published in 2019. At the end of the year 2019, the Chinese biophysicist He Jiankui, the medical father of the first ever germline CRISPR/Cas gene-edited babies, was sentenced to 3 years' imprisonment with a three-million-yuan fine. In 2020, Emmanuelle Charpentier and Jennifer Doudna were honoured with the Nobel Prize in Chemistry *for the development of a method for genome editing*. With this translation, the publisher and I wish to contribute to your understanding of the matter, its background, and ultimately your scientific literacy.

In the original book, I used the term "Gen-Schere", which translates best as genetic scissor but is rarely used in English. I have replaced it with either CRISPR/Cas system or gene editing.

Since 2019, many thousands of human genomes have been analysed, and artificial intelligences are associating their genetic make-up with traits such as disease or cognitive abilities. By the way: This book was translated by an artificial intelligence, a neural network to be precise, too. While proofreading the translation generated by DeepL, I was both amazed and sometimes amused. Once again, it showed me the importance of context, as with the genetic code. I tried my best to correct all artificial errors.

Finally, wherever possible, I replaced German references by English ones.

Mittweida, Germany   Röbbe Wünschiers

Contents

1 Preliminary Thoughts 1
   References. 7

2 What is Genetic Information? 9
   2.1   The DNA Molecule 11
   2.2   The Genetic Code 15
   2.3   The Gene 17
   2.4   The Genome. 20
   2.5   The Morphogenetic Code 23
   2.6   The Epigenetic Code 29
   References. 30

3 Breeding, Yesterday Until Today 35
   3.1   Nuclear Gardening. 41
   3.2   Conventional Breeding Methods 44
   3.3   Colourful Genetic Engineering 48
   3.4   Green Genetic Engineering 52
   3.5   Mottled Genetic Engineering. 67
   3.6   Organic Farming and Genetic Engineering 75
   3.7   Risks of Genetic Engineering 78
   References. 94

4 Reading Genetic Material 105
   4.1   Genetic Variation in Humans. 111
   4.2   Genetic Diagnostics. 117

    4.3   Prenatal Diagnostics. 123
    References. 136

5   Editing Genetic Material 141
    5.1   CRISPR/Cas System 142
    5.2   China's CRISPR Crisis? 156
    5.3   Gene Therapy. 165
    References. 171

6   Writing Genetic Material. 177
    6.1   Fabrication of Life 178
    6.2   Synthetic Biology. 181
    References. 195

7   Genes and Society 199
    7.1   Citizen Science. 200
    7.2   Commercializing Genetic Information 211
    7.3   My Genes and Me 218
    7.4   Gene Banks 225
    References. 230

8   Rethinking Genetics 237
    8.1   Epigenetics. 239
    8.2   Artificial Intelligence 247
    8.3   Dynamic Hereditary Material 252
    References. 257

9   Well Then?. 263
    Reference 265

Index. 267

1
Preliminary Thoughts

While the gene editing generation is settling the score with the fossil/plastic-waste/fine-dust generation in terms of climate and environmental protection, methods are emerging in the world's laboratories that could contribute both to climate and environmental protection—or make everything much worse—with genetic engineering knowledge and practices. At the latest since the Swedish schoolgirl Greta Thunberg drew the attention of young, but also older people to the environmental problems of our planet, it has finally become clear: we, young and old alike, have a long-term responsibility that we lived up to more badly than rightly. We have long since given our Earth age its own name: the **Anthropocene,** a term popularised by the German chemist and Nobel Prize winner Paul Crutzen and the US biologist Eugene Stoermer in 2000 [1]. The Anthropocene describes the current era in which humans have become a major influence on biological, geological and atmospheric processes. We are witnessing massive species extinctions, a retreat of permafrost and melting of glaciers and polar ice caps, and the formation of new sediment layers, including plastic particles. At the same time, we know that all living things on our planet are based on a

© Springer-Verlag GmbH Germany, part of Springer Nature 2022
R. Wünschiers, *Genes, Genomes and Society*,
https://doi.org/10.1007/978-3-662-64081-4_1

simple code consisting of only four building blocks: the genetic code, molecularly written down in DNA (Chap. 2). Can it help us?

Since the 1950s, the beginnings of molecular biology, we have begun to understand this code. Since the 1970s, the beginnings of genetic engineering, we have been able to specifically modify it. Since 1986 we are deliberately releasing genetically modified organisms (initially plants) and since then it can be said that our **ecological footprint** has been joined by a **genetic footprint.** The current genetic engineering revolution was heralded in 2012 when three scientists copied from nature a process for altering (editing) the genetic code with great precision. Imagine editing one of 3.2 billion letters—that is the number of building blocks in the human genome. This book, by the way, has about 480,000 characters. That is the precision of the **new gene editing** (Sect. 5.1), which has also found its way into contemporary German literature like in Martin Sutter's novel "*Elefant*" [2].

And then the "*genetic engineering hammer*", as the German tabloid BILD wrote: On November 26, 2018, the Chinese scientist Jiankui He announced in the course of a scientific conference that for the first time he had sustainably modified the genetic material of at least two babies, the twins Nana and Lulu [3]. Sustainable means that the offsprings of **Nana and Lulu** will also carry the genetic modification in every single cell. A taboo has been broken. In the Anthropocene, the Anthropo-gene, the man-made gene, is created. Where should and can we go from here?

Nana and Lulu are unintentionally sustainable representatives, but also products of the gene editing generation. By the gene editing generation, I primarily mean the currently living age cohort that still has the procreation of offspring ahead of it. This generation has not only an immense global responsibility with regard to the environment and the earth's climate, but also with regard to the genosphere [4]. This refers to the totality of all genetic systems that ensure the existence, regeneration and reproduction of the biosphere (Sect. 7.4). This generation will be confronted with the question of how much genetic engineering knowledge and genetic engineering practices they want to use, at the latest when they go to the gynaecologist while pregnant or when they go to reproductive physicians with a desire to have children. And those doctors and researchers who make gene editing available and develop them further are what I mean in second place as the gene editing generation. As a society, we are faced with the question of what means are acceptable to us in order to fulfil our responsibility towards the planet and future generations. Can, may or even must genetic engineering contribute to the solution?

The *framing of the* public discussion by the critics seemingly excludes this possibility (Fig. 1.1). Scientists find it harder than ever to be heard in today's world. Complex

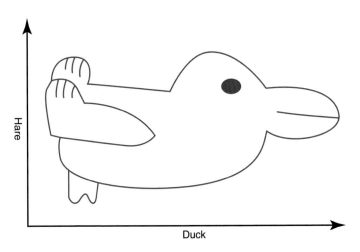

Fig. 1.1 It is all a question of perspective. Skilful framing can quickly turn a rabbit into a (newspaper) duck

discourses do not fit into our fast-moving times. And industry, with its monopolization and capitalization of genetic engineering, especially in the seed business, has contributed quite significantly to the current, albeit very vague, public opinion against genetic engineering. In addition, there is the European trauma of eugenics, which was first formulated in England, developed further in the USA and fatally abused in Nazi Germany. This mixture has given rise to today's fear that genetic engineering could escape democratic control through the power of capitalism. And thus, the discussion of risks (Sect. 3.7) of genetic engineering is usually less about the technology itself than about its social embeddedness. My argument is not that genetic engineering is the best of all solutions. But I do argue clearly against it being the primary culprit for problems such as the decline in biodiversity or the pollution of arable land with pesticides [5]. Sewage treatment plants have problems not because there are toilets, but because people dispose of their antibiotics in the toilet; washing machines are not to blame for environmentally unsound detergents; genetic engineering does not exempt us from good agricultural practices. I do, however, speak out against the prevailing sweeping actionism. All activities to combat climate change must be measured against their effect on the Earth system as a whole, as the English chemist James Lovelock and the US microbiologist Lynn Margulis formulated in the 1970s with their **Gaia hypothesis** [6].

When I told a child psychologist a few years ago about my intention to write a book about genetic engineering, she said: *Yes, that's an important topic.* And then, *I hope you're against genetic engineering.* This reflexive reaction against genetic engineering relaxed in further conversation after we had illuminated various aspects and discussed scenarios. But I often experience this: rejection as a reflex, illumination of the topic, differentiation of opinion. Cases of the use of genetic engineering are then assessed differently and differences between genetic engineering and genetic technology also become clear. The colourful picture does not necessarily make a decision for or against the use of genetic engineering, even in individual cases, any easier. But we have to take the time. This reflex also reinforced my need to deliver this book as a contribution to this debate.

In 2018, a study was published describing that **extreme opponents** of genetically modified food have a below-average level of education in genetic engineering [7]. In contrast, however, these same people consider themselves to be particularly well informed. This observation was made in Germany as well as in France and the USA. The study came to the same conclusion with regard to the attitude of the interviewees towards genetic engineering and their knowledge of the medical application of genetic engineering such as gene therapy (Sect. 5.3). The situation is different when it comes to the topic of climate change: representatives of extreme positions demonstrate greater factual competence here. The study thus once again underlines the **emotionality** of the topic of genetic engineering. And it shows the well-known phenomenon that extreme attitudes are often accompanied by a closing off to other and new information. I argue that this also applies to **extreme proponents.** This does not even have to be intentional, but can also run subconsciously. This phenomenon, known as an **anchoring effect**, is used by our brains to attach new information to existing information. For example, the idea that climate change is a real threat to humanity can lead us to associate any information about natural disasters with climate change. Applied to genetic engineering, this can lead to the **black-or-white fallacy** that we can only do with or without it. And I experience this again and again: In February 2016, I participated in a panel discussion on *Green genetic engineering: devil's work or ethical imperative?* .

During the discussion with experts and the audience, I urged caution in the use of the then still fairly new methods of gene editing with the CRISPR/Cas system—more of which later (Sect. 5.1). As is all too often the case, I once again experienced black-and-white painting. Neither supporters nor opponents were able to approach each other, the fronts were fixed. As if to confirm this, a former colleague from my time at *BASF Plant Science* approached me after the lecture and accused me of stirring up fears. She said that the new technology offered unimagined opportunities and that we had to be careful this time to inform the public well about the advantages instead of talking too much about the potential disadvantages and thus spreading unnecessary worries. Knowing the lady well, I knew what she meant. Contrast *this time* with the past: the introduction of genetic engineering into agriculture in the 1980s. At that time, not much was enlightened, but simply genetically feasible things were implemented. The broken trust, as mentioned, is interpreted today by many market experts as a source of resistance to genetic engineering. With the new genetic engineering, gene editing based on the CRISPR/Cas system, a new attempt could be made to publicly and controversially discuss the opportunities and risks of genetic engineering. By controversial, however, I do not mean a clash of fronts, as we have seen so far. I expect controversial discussions to take place within each front. We need to move away from black or white and (learn to) think more colourfully.

Finally, take a little test: Can you explain the meaning of *genome*? Numerous recent international studies show that most people are genetically illiterate [8, 9]. The word genome has no clear meaning for everybody and is most likely to be associated with gene manipulation. Genes are as abstract as atoms. And manipulating genes cannot bode well. After all, we do not like to be manipulated by the media. Genetic modification? That is more like it. Genetic optimization? An American social researcher laughed at me when I used that term: *You want to optimize nature?* She meant that nature is already optimal after all. Design? With gene editing we can design living beings. Hmm. Designing living beings on the drawing board and then constructing them in the lab?—I show you where we are with that in Sect. 6.2. So, let us not be manipulated and taken in by distorted images through contrived names! Let us take them for what they are: Meaningful images.

After these preliminary thoughts, now accompany me into the simply fascinating world of molecular biology and genetics and paint your own personal picture.

References

1. Crutzen PJ, Stoermer EF (2000) The "Anthropocene". Global Change Newsl 41: 17–18
2. Suter M (2018) Elefant. 4th Estate, London
3. Gentechnologie-Forscher: Designer-Babys in China geboren. (2018) In: Bild. Aufgerufen am 23.04.2019: bild.de/news/ ausland/news-ausland/gentechnologie-forscher-designerbabys-in-china-geboren-58647586.bild.html
4. Sauchanka UK (1997) The genosphere: The genetic system of the biosphere. Parthenon Publishing, Carnforth/UK
5. Klümper W, Qaim M (2014) A Meta-Analysis of the Impacts of Genetically Modified Crops. PLoS One 9: e111629. doi:https://doi.org/10.1371/journal.pone.0111629
6. Lovelock JE, Tellus LM (1974) Atmospheric homeostasis by and for the biosphere: the Gaia hypothesis. Tellus 26: 2–10. doi:https://doi.org/10.3402/tellusa.v26i1-2.9731
7. Fernbach PM, Light N, Scott SE, et al (2018) Extreme opponents of genetically modified foods know the least but think they know the most. Nat Hum Behav 3: 251–256. doi:https://doi.org/10.1038/s41562-018-0520-3
8. Middleton A, Niemiec E, Prainsack B, et al (2018) "Your DNA, Your Say": Global survey gathering attitudes toward

genomics: design, delivery and methods. Pers Med 15: 311–318. doi:https://doi.org/10.2217/pme-2018-0032
9. Boersma R, Gremmen B (2018) Genomics? That is probably GM! The impact a name can have on the interpretation of a technology. Life Sci Soc Policy 14: 8. doi:https://doi.org/10.1186/s40504-018-0072-3

Further Reading

Kay LE (2000) Who wrote the book of life? A history of the genetic code. Stanford University Press, Stanford.
Knoepfler P (2016) GMO sapiens: The life-changing science of designer babies. World Scientific, New Jersey. doi:https://doi.org/10.1007/978-3-662-56,001-3
Meloni M (2016) Political Biology. Palgrave Macmillan, Hampshire/UK. doi:https://doi.org/10.1057/9781137377722

2
What is Genetic Information?

We live in the age of DNA, the deoxyribonucleic acid. This is the complicated chemical name for the carrier molecule of genetic information, which contains the blueprint of every cell from bacteria to humans. In German one writes actually **DNS**, whereby the S stands for Säure. In France DNA is called ADN (*acide désoxyribonucléique*). In 2018, **DNA** has been officially added as emoji, which can graphically enrich tweets, posts and other messages (Fig. 2.1).

Before a cell divides into two daughter cells, it duplicates its genome. We humans pass on our genetic information to the next generation via egg and sperm cells, the so-called germ cells. This can lead to changes (mutations) in the information. These arise either when the information is copied or as a result of exposure to chemicals or radiation (Sect. 3.1). Mutations contribute to the fact that no living being is like another. Even identical twins have been shown to differ in their genetic information, although only minimally [1]. Mutations can be harmful, beneficial or have no effect, i.e., they can be neutral. It is the variability of the genome that drives evolution through variation and

© Springer-Verlag GmbH Germany, part of Springer Nature 2022
R. Wünschiers, *Genes, Genomes and Society*,
https://doi.org/10.1007/978-3-662-64081-4_2

Fig. 2.1 In 2018, 65 years after its structure was elucidated, DNA and also viruses became available as emoji

selection, as described by Charles Darwin and Alfred Russel Wallace in the mid-nineteenth century. For some time now, however, we have also known that experiences in the widest sense, gained during one's lifetime can be passed on to subsequent generations, as Jean-Baptiste Lamarck also assumed at the beginning of the nineteenth century. The mechanism behind this is known as epigenetics and is currently revolutionising thinking about medicine, genetic engineering and evolution. It is described in more detail in Sect. 8.1.

2.1   The DNA Molecule

Sometimes I wonder whether the term DNA needs to be explained at all, since it seems to have arrived in society. BMW boss Harald Krüger, for example, speaks of *corporate DNA* [2] in the field of vehicle construction. In an article about the Catholic Church I read about *Catholic DNA*, [3] and in a report about the English Kingdom and the Brexit I read about *cultural DNA* [4]. The German philosopher and journalist Thorsten Jantschek, in a video message on the occasion of the awarding of the Prize of the Leipzig Book Fair, even talks about the fact that the book *corresponds to the spiritual DNA of the Republic* [5]. DNA is a symbol for something common and meaningful. Well, DNA is indeed common to all forms of life, and as genetic information it also provides meaning.

For most genetic engineers, DNA is reduced to the sequence of the four letters A, T, G and C. So: ...CGATTAGCTGCT... A, T, G and C are the abbreviations for the **nucleobases** adenine, thymine, guanine and cytosine. Together with the sugars ribose and phosphate they form the building blocks of DNA, the **nucleotides** (Fig. 2.2).

They form, like a string of pearls linked together, the hereditary molecule DNA. This molecule forms a **double helix,** i.e., consists of two molecular strands. The detailed structure was elucidated in 1952 by the English biophysicist Francis Crick and the American geneticist James Watson on the basis of X-rays taken by the English chemist Rosalind Franklin. They found that the two strands were complementary: If the sequence of nucleotides (DNA sequence) of one strand is known, then the sequence of the opposite strand can be elucidated, since A always pairs with T and G with C (**base pairing** of complementary nucleotides. The

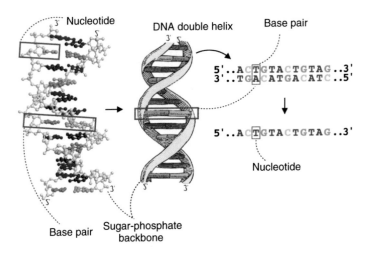

Fig. 2.2 Different representations of DNA with twelve base pairs as a molecular model (left), schematic structure (middle), text with both single strands (top right) and one single strand (bottom right). According to the chemical nature of DNA, it is assigned a direction with the designations 5′ and 3′

human genome consists of 3.2 billion nucleotides distributed over 23 **chromosomes** of different lengths (Fig. 2.3). One can also speak of 3.2 billion base pairs (**bp**) instead of 3.2 billion nucleotides (**nt**). In addition, there are 16,569 nucleotides each on the chromosome of the **mitochondria,** the cells' energy power stations.

In most animals, plants and humans, the chromosomes are present as double copy (**diploid**), sometimes even multiple copied (**polyploid**)—in bacteria, on the other hand, they are usually single copied (**haploid**). Thus, every human cell has inherited 23 **chromosomes** each from the father and mother. If the DNA of the 46 chromosomes of a single human cell were combined to form a thread, it would be about two metres long. The DNA of all human cells would reach from the earth to the sun and back about four

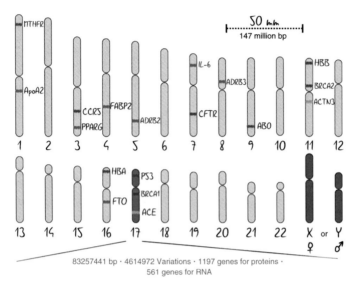

50 mm
|··|
147 million bp

83257441 bp · 4614972 Variations · 1197 genes for proteins ·
561 genes for RNA

A total of around 3.2 billion bp with over 360 million variations
Mitochondrium: 16,569 bp · 0.0054 mm

Fig. 2.3 The simple (haploid) human chromosome set. The chromosomes are between 16 and 85 centimetres long. CFTR refers to the location of the gene that causes cystic fibrosis (mucoviscidosis) in a defective form. AB0 is the location of the gene that determines blood groups. The CCR5 gene is associated with resistance to the HI virus. The genes HBA and HBB are mutated in α and β thalassemia, respectively. At blue marked positions are genes that the company *for me do* uses for the classification of nutrition types; at orange positions, however, are genes that allow predictions about the type of athlete (Sect. 7.3)

times (nine billion kilometres). That is just about the same as the orbit of the planet **Saturn** around the sun. The Japanese berry *(Paris japonica)* has a genome size of around 149 billion nucleotides distributed over ten chromosomes, making it almost 47 times larger than the human genome and would form a continuous DNA thread 91 m long [6]. Directly after this comes the marbled lungfish *(Protopterus* *aethiopicus*) with a genome size of 130 billion nucleotides distributed over 14 chromosomes. The smallest known genome with only 159,662 nucleotides is that of the bacterium *Carsonella ruddii,* which lives symbiotically in leaf fleas.

On the chromosomes, the genetic information is distributed among **genes** (Sect. 2.3). The totality of all genes of a living being is called **genome** or **genotype.** A gene can be understood as a package of genetic information that codes for a specific trait, for example the blood group. The totality of all characteristics of a living being makes up its **phenotype**, its appearance. For example, since there are several blood groups (0, A, B, AB), there must be several variants of the gene, which we call **alleles** (Fig. 4.6). From each gene we carry one maternal and one paternal allele. Some traits, such as eye colour, involve at least eight genes [7].

The basis of all diagnostic analysis and genetic engineering work is a thorough understanding of the function of a particular section of the genome. In the early days, this was only possible in a very rough way. So-called genetic **markers** were associated with phenotypic manifestations such as diseases or other characteristics. These markers were initially not nucleotide sequences (DNA sequences), but rather physical observations, such as the fact that DNA breaks down into fragments of varying size after treatment with a DNA-cutting enzyme (restriction enzyme). The size and distribution of the fragments could be measured and correlated with characteristics. Today we can read the entire genetic material (the **DNA sequence**) of a living being, from bacteria to humans (Chap. 4). Approximately 99.5% of a person's genome is similar nucleotide by nucleotide (base pair by base pair) to the genome of any unrelated other person (Sect. 4.1) [8, 9].

2.2 The Genetic Code

How can the genetic information in the form of the sequence of 3.2 billion nucleotides contain the blueprint of cells, even of entire living beings? To understand this, we must consider the flow of information (Fig. 2.4).

Analogous to words in a text, there are strings of letters that are translated into **proteins**. A certain class of proteins are **enzymes**. They form the toolbox of a cell, because they are responsible for the metabolism. Another class of proteins are building materials, such as the keratin from which

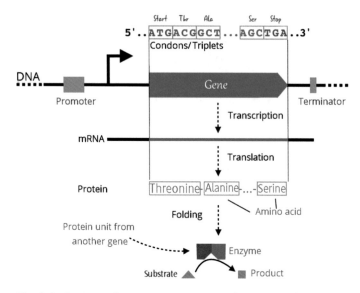

Fig. 2.4 Processes in gene expression. Codons on the hereditary DNA molecule code for amino acids, the building blocks of proteins. As a rule, proteins consist of several hundred amino acids. The promoter region of the DNA serves to regulate transcription. A protein can have an enzymatic activity, either alone or, as shown, in concert with other proteins and thus convert chemical substances (substrates)

our hair is made. Proteins, like DNA, are also made up of a chain of molecular building blocks, the **amino acids**. The sequence of three nucleotides on the DNA (a so-called **codon** or **triplet**) codes for one amino acid. A total of twenty amino acids are coded in this way (Sect. 6.2). In addition, there are codons that mark the start and end of the protein. If the DNA contains the sequence ...ACGGCT...AGC... in the protein, this translates into the amino acid sequence ...-threonine-alanine-...-serine-... Thus, this process is called **translation**. It is preceded by a **transcription** of the DNA information into an RNA molecule, the so-called *messenger* RNA (mRNA). **RNA** (ribonucleic acid) is single-stranded, differs slightly in its chemical structure from DNA, is therefore more mobile in the cell and can be broken down more quickly. The process that expresses the genetic information stored on the DNA via the RNA to the protein is called **gene expression**. A small change in the DNA sequence, such as a nucleotide exchange in the codon ACG to CCG, can thus lead to a change in the protein (threonine in proline). And since a protein fulfils a function, functional changes or failures may be the consequence. In this way, around 20,000 proteins are encoded in the human genome.

The fact that the genetic code applies equally to all known living beings led to the famous saying of the French biochemist and Nobel Prize winner Jacques Monod that everything that applies to bacteria must also apply to elephants [10] (Fig. 2.5).

Interestingly, only about 3% of our DNA codes for proteins. What is the rest for? In order to answer this question, we need to take a detailed look at the structure of genetic information.

Fig. 2.5   (Almost) everything that applies to the bacterium *Escherichia coli* will also apply to an elephant

2.3   The Gene

The term gene (Sect. 2.1) obviously plays a central role in genetics, genomics and genetic engineering. It also seems that, just like DNA, genes have taken their place in our vocabulary. Examples of this may be a newspaper article entitled *Transporter with car genes* [11] in the section on cars and traffic, or the placing of the term *Sarrazin gene* in third place among the words of the year 2010 in Germany, after *Wutbürger* and *Stuttgart 21* [12]. In fact, however, hardly any other term in genetics is as much in discussion as the gene [13]. Over the years, the view on the nature of genes has changed. In his 1976 book *The Selfish Gene*, Richard Dawkins attributed to the gene a selfish end in itself, loosely based on the motto: The chicken is only the transition from one egg to another [14]. The background is the fact that at least individuals with sexual reproduction pass on only half of their genes to the next generation. According to this, there is competition between the genes for passing them on. Genes that are not alleles and therefore do not compete with each other can therefore also cooperate. Itai Yahnai and Martin Lercher outline cooperation as a counter-proposal in their book *The Society of Genes*: A gene alone cannot achieve anything, nor are we merely the sum of our genes [15]. Even a simple bacterium can only develop and live through the ordered interaction of many genes or their products such as proteins or RNA molecules.

Recent findings show that a set of about 470 genes is necessary to "run" a living and reproducing cell [16, 17]. Whether this number can be even smaller depends largely on how much dependence on external factors the cell with the minimal genome is allowed to have. That research into the smallest possible set of genes has practical significance can be seen in Sect. 6.2. In his book *Treffen sich zwei Gene* (*Two Genes Meet*) from 2017, Ernst-Peter Fischer makes it clear that the term gene cannot be defined precisely at all [18]. He thus draws a parallel to the atom, which must be seen less as a part than as a model. And that is a good point: the gene can be seen as a metaphor for a functional unit in our genome, encoded in our DNA.

The term gene was coined in 1909 by the Danish botanist Wilhelm Johannsen. He referred to all the objects that the theory of heredity deals with as descendants, using the Greek noun *genos*. For Johannsen, however, they were only a mathematical quantity. Three years earlier, the English biologist William Bateson had described the science of heredity, after the Greek adjective *gennetikos* for producing, as **genetics.** That genes consist of matter and are encoded on DNA was still unthinkable at that time. A common definition describes the **gene** as a section of DNA that codes for a protein (Fig. 2.4). From this, it can be deduced that humans have about 20,000 genes. However, it is important to know that a protein-coding DNA section is not always active, i.e., it is not always expressed. A nerve cell contains different proteins than a liver cell. And bacteria, for example, need other proteins, especially enzymes, if they feed on lactose instead of glucose. Genes, more precisely gene expression, are therefore regulated. And this is done via DNA sections in front (the **promoters**) and behind (the **terminators**), the actual protein-coding regions (Fig. 2.4). These regulatory regions do not code for a protein, but they do have an important function. Changes (mutations) therefore often lead to altered expression and behaviour of the cell. And the remaining sections? For many years, these DNA regions were called *garbage DNA* until it was observed that information was also encoded there. Thus, there are numerous DNA sections that encode regulatory RNA molecules, but which are not translated into proteins. That is why today we speak more cautiously of *junk DNA*: junk is kept and garbage is thrown away. But even if we take all these DNA sections for genes including regulatory regions and regulatory RNA together, most of the DNA is still noncoding. But even if their function is still unknown, at least they can make a structural contribution to gene regulation. In order to do so, we have to remember that although the DNA thread is present in the cell as a bulge, it certainly plays an important role as to which areas are near or far away.

Since we humans have inherited one maternal and one paternal copy (**allele**) of each gene, both contribute to the phenotype (Sect. 2.1). If one copy is defective, the other copy can take over the function. In this case, the defective gene is **recessive** (masked). If one copy codes for a cytotoxin, for example, then the effect is **dominant** and outshines the effect of the second allele (Fig. 4.6). Diseases that are inherited recessively therefore only occur when both the maternal and paternal allele are defective in a person. However, the recessive mode of action of the genes only affects the autosomes, that is chromosomes 1 to 22 and in women the X chromosome (sex chromosomes). Only these are present in two copies. In men, both sex chromosomes are present in a single copy. This means that a recessive X or Y gene automatically has a dominant effect in males, since there is no intact replacement allele in the cell.

As we will see in the following section, not all genes lie peacefully in the genome: there are so-called **jumping genes.** These contribute significantly to the difficulty of defining genes, because actually—actually—they lie dormant in the genome and seem to have no function. However, they make up a large part of the junk DNA in the genome.

2.4   The Genome

Due to technical progress in the analysis of the DNA sequence (Chap. 4), the term gene is increasingly being replaced by the word genome. The genome refers to the totality of all genes of a living being (Sect. 8.3). Genetics, which focuses in particular on the effect and inheritance of individual genes, is increasingly becoming **genomics,** which focuses on all genes. The human genome consists of the 46 chromosomes and the chromosome in the mitochondrion (Fig. 2.3). Of the 46 chromosomes, 44 are **autosomes,** the maternal and paternal chromosomes 1 to 22. They are joined by the sex-determining X and Y chromosomes, the **gonosomes** or sex chromosomes. More precisely, the genome consists of 44 autosomes, two gonosomes and one mitochondrial chromosome (also called **mtDNA**). We pass half of our genome on to our descendants, or more precisely, men pass 23 chromosomes and women 23 chromosomes plus the chromosome of the mitochondria. Since the mitochondrial genome in sperm is lost during egg fertilisation, only the mother's mtDNA is passed on [19]. An exciting side note: Since October 2018, a research consortium with scientists from the USA, China and Taiwan has been questioning this dogma of heredity, which has led to a tangible scientific dispute [20–23]. The end is still open.

Since the small chromosome of the **mitochondria** is very easy to isolate and sequence, it has become the gold standard of parentage research. As a result, only the maternal line was traced, which is why the search for the biological **mother** was launched, who was baptized Eva [24]. Every living person has inherited his or her energy-releasing mitochondria from this woman, who lived about 150,000 years ago, or about 6000 generations. This is why she is also known as a *lucky mother*, who was able to pass on her genetic material in such a sustainable way.

Why do we only pass on one set of chromosomes to our offspring? Well, strictly speaking, we pass on a mix of our paternal and maternal chromosomes, thanks to the birth of sexual reproduction (Fig. 2.6). In asexual reproduction, the genetic material, the chromosomes, are simply duplicated and divided between two daughter cells. Apart from small

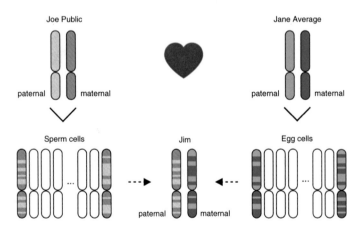

Fig. 2.6 In sexual reproduction, the chromosome sets are mixed during the maturation of egg and sperm cells. Only one chromosome is shown at a time

mutations that occur during copying, both copies are identical. Asexual reproduction, for example, is the basis for the growth and renewal of cells in animals and plants.

But we already know about sexual reproduction from bacteria, where genetic information is exchanged between sexual partners. This results in a mixing (**recombination**) of the parental genomes, or chromosomes. The best-known process for this is the *crossing-over* during the formation of the **germ cells**, that is the egg and sperm cells (Fig. 2.6). This increases the genetic variability between individuals and thus their adaptability. When **cloning** a living being, such as Dolly the sheep in 1996, the English biologist Keith Campbell and the English embryologist Ian Wilmut succeeded in creating identical copies of the diploid genomes and thus of the living beings (Sect. 3.5 and Fig. 3.12) [25].

Since the genomes of all individuals of a species are different, only a fraction can be seen when the genome of an individual is decoded. Thus, although the sequencing of the human genome was a breakthrough (Chap. 4), it was initially only from individual genomes. However, in order to get to know as many gene variants as possible, many individuals must be sequenced. The genomes of all individuals of a species are called **pangenomes**. Certainly, it is hardly possible to sequence all individuals. But projects in this direction are underway. In the international *1000 Genomes Project,* the USA, England, China and Germany have been analysing the diversity of the human genome in a joint effort since January 2008 (Sect. 4.1). As of March 2019, the genomes of over 2500 people have been sequenced and analysed in this project [26]. These data are freely available. The situation is different in the project of the Icelandic company deCODE Genetics [27]. The genome data of more than 2600 Icelanders belong to the company and are exclusively sold (Sect. 7.2). A similar situation is likely to occur in the largest sequencing project to date: The *National Health and Medicine Big Data Nanjing Center* in the Chinese province of Jiangsu, announced in October 2017 that it intends to sequence the genome of one million Chinese. This clearly shows that there is great interest in exploring all genetic variants of our pangenome. In combination with medical records, they can be used for diagnostic purposes and therapeutic strategies (Sects. 4.2 and 8.2).

2.5   The Morphogenetic Code

Famous is the book by the German biologist and Jena professor Ernst Haeckels entitled *Kunstformen der Natur* [28] (Art Forms in Nature). The first volume was published in 1899 and contains impressive lithographs of animals and plants (Fig. 2.7). Even today, you will hardly find a modern antiquarian bookshop that does not offer reprints of Haeckel's volumes on display. How does this diversity come about? In his book *The sculpture of life*, published in 1973, the Hungarian-American microbiologist Ernest Borek asks why living beings do not look like spaghetti when the genetic code is linear like a thread [29].

It is one of nature's greatest wonders how a sequence of four nucleotides can give rise to a complex creature with an incredible range of forms. Personally, I am filled with a mixture of humility, amazement and curiosity, which is not diminished by the mathematical derivation of the origin of the giraffe skin pattern from chemical concentration gradients [30]. The US-American biologist and anthroposophist Craig Holdrege describes the mode of action of the hereditary molecule as follows:

> The relationship between forest and seed is comparable to the mutual dependence of organism and DNA. The DNA is dependent on the organism as an environment and at the same time directs the abilities and characteristics of the organism in a certain direction [31].

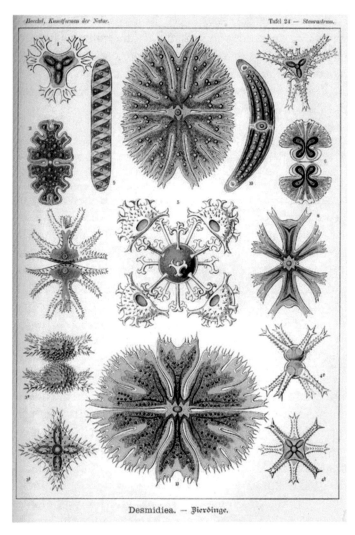

Fig. 2.7 Green algae of the order Desmidiales—such algae swim in our puddles. How do the forms of the cell walls and cell components of these unicellular organisms develop? (Source: Haeckel [28])

I would just like to give a tiny insight into how form can be created from genetic information, so that we do not lose sight of the big picture [32].

We have already learned about genes and seen that the activity of genes is regulated by promoters (Fig. 2.4). Promoters are DNA sequences ahead of genes. They can have a very complex structure and contain several DNA sequences as binding sites that are recognized by specific proteins—we call these **transcription factors** because they regulate transcription. When a protein binds to the DNA, this influences the regulation of gene activity. Genes can thus be activated, but also inactivated. In some cases, there are several binding sites for the same protein. All binding sites must then be occupied, for example to activate the gene (Fig. 2.8). This can be thought of as a logical AND circuit or as a series connection of switches in an electric circuit. To be able to occupy all binding sites, the protein must be present in a certain minimum quantity (minimum

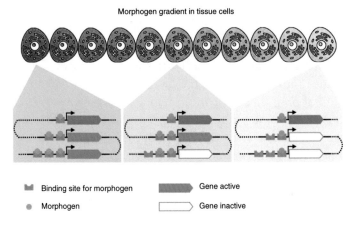

Morphogen gradient in tissue cells

Binding site for morphogen   Gene active

Morphogen   Gene inactive

Fig. 2.8   A concentration gradient of a morphogen (for example a protein) in a tissue can lead to a concentration-dependent activation of genes. For this, all binding sites must be occupied

concentration). We now imagine that the binding protein, the transcription factor, can diffuse (migrate) from cell to cell—just like sugar in a tea glass, which spreads slowly even without stirring. If the transcription factor is produced at one end of a tissue, it diffuses from cell to cell, but its concentration decreases. If we now combine the concentration gradient of a DNA binding protein and a variable number of DNA binding sites in the promoter region of genes, then we have an example of how site-specific genes can be regulated. In the example in Fig. 2.8, we see three genes that have different sensitivity to the transcription factor.

The sensitivity is regulated by the number of binding sites. We call a transcription factor that is involved in shaping a protein a morphogen—not to be confused with the gene. **Morphogens** are not sections of DNA, but signalling substances in cells and tissues. The morphogen can be a protein, but also another chemical substance, for example a hormone. Physical parameters such as temperature or pressure are also possible, although they must be converted back into a chemical signal in some way. We know this from our light-, pressure- or temperature-sensitive nerve cells (sensor neurons).

The **morphogenetic code** determines the formation of tissues and living beings, the morphogenesis. It contains a kind of position information. And already even a small cell cluster can have a direction, a front and a back. Through the interaction of several morphogens and a diverse network, the formation of complex developmental structures and forms is thus possible: a fern, a fish, man. Diversity in networking can be achieved, for example, with AND circuits, and additionally with OR circuits, so that one gene reacts to several morphogens. Some genes can also be activated, while other genes can be inactivated by the same morphogen. An activated gene can also form another morphogen. And so on. I know: it is still a long way from the model described above to a complete human being or even just a shape-rich alga (Fig. 2.7). But I hope that the working principles has become apparent.

It is reasonable to assume that greater complexity is based on more genetic information. This can be measured approximately. Even before it was possible to sequence the genetic material, it was possible to isolate DNA from cells and weigh it. The **C-value** is the name for the amount of DNA in a cell, measured in grams—today in base pairs—and corrected to a simple set of chromosomes. This gives us the amount of genetic information. And the complexity? Is a turtle more complex than a starfish? As a measure of **complexity,** we can take the number of different cell types of a species. We have liver cells, kidney cells, skin cells and many more—about 200 cell types are attributed to mammals. Single-celled organisms, on the other hand—have one. In addition, we can add the evolutionary stage of development, since a reduction of complexity is complex. Here the inconspicuous grasses should be mentioned, which are evolutionarily much more advanced than the most beautiful flowering plants. If we now relate complexity to the size of the genome, it quickly becomes clear that the complexity of a species cannot be derived exclusively from the **size of the genome** (Fig. 2.9). This observation is also known as the **C-value paradox** [33]. Since the 1980s, however, this paradox has been largely resolved. Greater complexity is achieved through more regulation and the recombination of existing genetic control loops.

With the latest sequencing methods (Chap. 4), not only the largely static genetic material, DNA, can be analysed. Its transcribed and thus active parts, such as mRNA (Fig. 2.4), can also be identified and quantified in this way. This makes it possible to show when which gene is active.

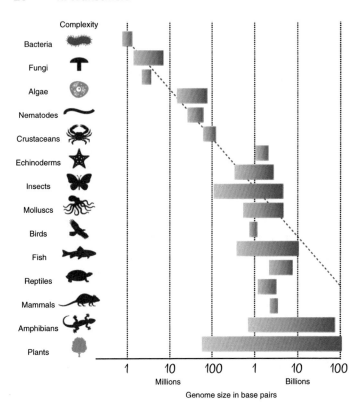

Fig. 2.9 The C-value paradox, according to which the complexity of living organisms is not consistently correlated with genome size

The latest developments even make it possible to do this for individual cells in a tissue. Recently, a research team led by Professor Marja Timmermans from the University of Tübingen used this method to track the root growth of a plant [34]. Around 15,000 transcripts (mRNAs) were found in 4727 cells. The respective active genes in a cell determine its so-called expression profile. It is like a genetic fingerprint, but it can change over time—for example, when the cells migrate from the root interior to the periphery and take over their final function. In this way, the researchers were also able to describe the spatial and temporal distribution of morphogens.

2.6 The Epigenetic Code

Many people know the rule of thumb that our living conditions are determined by genes and the environment to about half. This rule probably holds more hope than research. There is no question that the environment plays an important role, a very important role indeed. But such a separation is questionable, since every environmental signal must hit a receiver in order to have an effect on a living being. And this receiver, be it a receptor molecule, a neuron, a tissue or an organ, has its origin in the genetic information. What is clear is that we inherit our genetic material from our ancestors. We will see how far the **genetic memory** extends when we look at our mitochondrial Eva in Sect. 2.4. It spans about 150,000 years or the equivalent of about 6000 generations. But an exciting question is: How close does it reach? That one half of our genetic make-up comes from each of our father and mother is—still—undisputed and has already been described in a poem by Johann Wolfgang von Goethe from 1827, set down in his *Zahmen Xenien* (*Tame Xenia*):

> My father gave my stature tall,
> And rule of life decorous;
> My mother my nature genial
> And joy in making stories;
> Full well my grandsire loved the fair,
> A tendency that lingers;
> My grandam gold and gems so rare,
> An itch still in the fingers.

If no part from this complex all
Can now be separated,
What can you name original
That is in me created?

We will learn about exceptions later. But Goethe addresses something interesting in his poem: What is peculiar to us? Only the recombination of the parental inheritance, as shown in Fig. 2.6? The latest findings show that the genetic code is extended by an epigenetic code over the genes (the Greek prefix *epi-* stands for up, over) [35]. This is a chemical change in the DNA caused by the **environment** and passed on to the offspring [36]. These findings are currently leading to a radical rethink about the effect of environmental factors such as fear, nutrition or sport on our genetic information (Sect. 8.1).

References

1. Weber-Lehmann J, Schilling E, Gradl G, et al (2014) Finding the needle in the haystack: Differentiating "identical" twins in paternity testing and forensics by ultra-deep next generation sequencing. Forensic Sci Int: Genet 9: 42–46. doi:https://doi.org/10.1016/j.fsigen.2013.10.015
2. Heuser UJ, Tatje C (2019) Danke, Diesel. Die Zeit: 11/2019, S. 17
3. Löbbert R (2019) Meine innere Kirche brennt. Die Zeit: 10/2019, Beilage Christ & Welt, S. 1
4. Bittner J (2019) Brexit: Im Tollhouse. Die Zeit Ausgabe 5, S. 3
5. Leipziger Messe GmbH (2019) Preis der Leipziger Buchmesse 2019. In: YouTube. Visited 21.03.2019: youtu.be/YZuadUiFtYo bei Minute 30
6. Pellicer J, Fay MF, Leitch IJ (2010) The largest eukaryotic genome of them all? Bot J Linn Soc 164: 10–15. doi:https://doi.org/10.1111/j.1095-8339.2010.01072.x

7. Liu F, van Duijn K, Vingerling JR, et al (2009) Eye color and the prediction of complex phenotypes from genotypes. Curr Biol 19: R192–R193. doi:https://doi.org/10.1016/j.cub.2009.01.027

8. The 1000 Genomes Project Consortium (2012) An integrated map of genetic variation from 1,092 human genomes. Nature 491: 56–65. doi:https://doi.org/10.1038/nature11632

9. The 1000 Genomes Project Consortium (2015) A global reference for human genetic variation. Nature 526: 68–74. doi:https://doi.org/10.1038/nature15393

10. Monod J, Jacob F (1961) Teleonomic mechanisms in cellular metabolism, growth, and differentiation. Cold Spring Harb Symp Quant Biol 26: 389–401. doi:https://doi.org/10.1101/sqb.1961.026.01.048

11. Elfer M (2018) Transporter mit Pkw-Genen. Mitteldeutsche Zeitung, Rubrik Auto & Verkehr, 29/30 Dezember 2018; S. 1

12. Redaktion (2010) Wort des Jahres 2010: Der „Wutbürger" sticht alle aus. In: Frankfurter Allgemeine Zeitung. Visited 14.04.2019: faz.net/1.581941

13. Pearson H (2006) Genetics: what is a gene? Nature 441: 398–401. doi:https://doi.org/10.1038/441398a

14. Dawkins R (2006) The Selfish Gene. Oxford University Press, New York/USA

15. Yanai I, Martin L (2016) The Society of Genes. Harvard University Press, Cambridge Massachusetts/USA

16. Juhas M, Eberl L, Glass JI (2011) Essence of life: essential genes of minimal genomes. Trends Cell Biol 21: 562–568. doi:https://doi.org/10.1016/j.tcb.2011.07.005

17. Hutchison CA, Chuang R-Y, Noskov VN, et al (2016) Design and synthesis of a minimal bacterial genome. Science 351: aad6253. doi:https://doi.org/10.1126/science.aad6253

18. Fischer EP (2017) Treffen sich zwei Gene. Siedler Verlag, München

19. Hutchison CA, Newbold JE, Potter SS, Edgell MH (1974) Maternal inheritance of mammalian mitochondrial DNA. Nature 251: 536–538. doi:https://doi.org/10.1038/251536a0

20. Luo S, Valencia CA, Zhang J, et al (2018) Biparental Inheritance of Mitochondrial DNA in Humans. Proc Natl Acad Sci USA 115: 13039–13044. doi:https://doi.org/10.1073/pnas.1810946115

21. McWilliams TG, Suomalainen A (2019) Mitochondrial DNA can be inherited from fathers, not just mothers. Nature 565: 296–297. doi:https://doi.org/10.1038/d41586-019-00093-1

22. Lutz-Bonengel S, Parson W (2019) No further evidence for paternal leakage of mitochondrial DNA in humans yet. Proc Natl Acad Sci USA 116: 1821–1822. doi:https://doi.org/10.1073/pnas.1820533116

23. Luo S, Valencia CA, Zhang J, et al (2019) Reply to Lutz-Bonengel et al.: Biparental mtDNA transmission is unlikely to be the result of nuclear mitochondrial DNA segments. Proc Natl Acad Sci USA 116: 1823–1824. doi:https://doi.org/10.1073/pnas.1821357116

24. Sykes B (2002) The Seven Daughters of Eve. Corgi, London UK

25. Wilmut I, Campbell K, Tudge C (2001) Dolly. Carl Hanser Verlag, München

26. Sudmant PH, Rausch T, Gardner EJ, et al (2015) An integrated map of structural variation in 2,504 human genomes. Nature 526: 75–81. doi:https://doi.org/10.1038/nature15394

27. Gudbjartsson DF, Helgason H, Gudjonsson SA, et al (2015) Large-scale whole-genome sequencing of the Icelandic population. Nat Genet 47: 435–444. doi:https://doi.org/10.1038/ng.3247

28. Haeckel E (1899) Kunstformen der Natur. Verlag des Bibiographischen Instituts, Leipzig und Wien

29. Borek E (1973) The sculpture of life. Columbia University Press, New York/USA

30. Turing AM (1952) The chemical basis of morphogenesis. Philos Trans R Soc, B 237: 37–72. doi:https://doi.org/10.1098/rstb.1952.0012

31. Holdrege C (1999) Der vergessene Kontext: Entwurf einer ganzheitlichen Genetik. Verlag Freies Geistesleben, Stuttgart

32. Carroll SB (2005) Endless Forms Most Beautiful. W. W. Norton & Company, New York/USA

33. Gregory TR (Edt) (2011) The Evolution of the Genome. Elsevier Academic Press, Burlington, Massachusetts/USA. doi:https://doi.org/10.1016/B978-0-12-301463-4.X5000-1

34. Denyer T, Ma X, Klesen S, et al (2019) Spatiotemporal Developmental Trajectories in the Arabidopsis Root Revealed Using High-Throughput Single-Cell RNA Sequencing. Developmental Cell 48: 840–852.e5. doi:https://doi.org/10.1016/j.devcel.2019.02.022

35. Deans C, Maggert KA (2015) What Do You Mean, "Epigenetic"? Genetics 199: 887–896. doi:https://doi.org/10.1534/genetics.114.173492

36. Tucci V, Isles AR, Kelsey G, et al (2019) Genomic Imprinting and Physiological Processes in Mammals. Cell 176: 952–965. doi:https://doi.org/10.1016/j.cell.2019.01.043

Further Reading

Carrol SB (2005) Endless Forms Most Beautiful. W. W. Norton & Company, New York/USA.

Fontdevila A (2011) The Dynamic Genome. Oxford University Press, New York/USA.

Mukherjee S (2016) The Gene: An intimate history. Thorndike Press, Farmington Hills/USA.

3
Breeding, Yesterday Until Today

[...] We have also large and various orchards and gardens, wherein we do not so much respect beauty as variety of ground and soil, proper for diverse trees and herbs, and some very spacious, where trees and berries are set, whereof we make divers kinds of drinks, beside the vineyards. In these we practise likewise all conclusions of grafting, and inoculating, as well of wild-trees as fruit-trees, which produce many effects. And we make by art, in the same orchards and gardens, trees and flowers, to come earlier or later than their seasons, and to come up and bear more speedily than by their natural course they do. We make them also by art greater much than their nature; and their fruit greater and sweeter, and of differing taste, smell, colour, and figure, from their nature. And many of them we so order as that they become of medicinal use. We have also means to make divers plants rise by mixtures of earths without seeds, and likewise to make divers new plants, differing from the vulgar, and to make one tree or plant turn into another. [...] Neither do we this by chance, but we know beforehand of what matter and commixture, what kind of those creatures will arise. [...] [1]

This text was written by the English philosopher and jurist Sir Francis Bacon in 1623. After his death in 1627, it was

© Springer-Verlag GmbH Germany, part of Springer Nature 2022
R. Wünschiers, *Genes, Genomes and Society*,
https://doi.org/10.1007/978-3-662-64081-4_3

published under the title *The New Atlantis*. It describes Bacon's utopian design of a perfect state, an optimal community, an ideal society. Science occupies a central place here, and so the quoted passage describes what animal and plant breeding is intensively researching today: precision breeding (Fig. 3.1).

Today, the term breeding implies a generational work of directing the reproduction of animals or plants. It can be assumed that with the transition from a hunter-gatherer way of life to sedentism at the transition from the **Ice Age** to the **Neolithic period** around 12,000 years ago, humans began to cultivate plants and keep animals in the broadest sense. One cannot speak here of breeding in the narrower sense, but nevertheless of a rather unconscious intervention in the gene pool. The **gene pool** describes the totality of all genes in a collection of organisms (population) that can reproduce among each other. In a figurative sense we could say that every human being has a mobile phone and the totality of all phones forms the mobile phone pool. If, for example, people in the Neolithic period cultivated emmer wheat or einkorn wheat, nurtured and cared for the growing plants and re-sown some of the harvested seeds, then the gene pool was reduced to the re-sown seeds. As a result of replanting this small subpopulation, inbreeding increasingly occurs—with consequences: Some traits become prominent. Often this has disadvantageous consequences and probably led to the fact that "fresh" genetic material, that is new seeds had to be collected and cultivated. However, advantageous traits could also emerge and be selectively cultivated further. Something similar can be thought of for livestock farming. In the beginning, therefore, "breeding" was probably limited to selecting plants that looked healthy and produced a good usable yield. We are talking here about **selection breeding**), which is directed at appearance and perceptible characteristics such as

Fig. 3.1 The highly developed island Bensalem from Sir Francis Bacon's future novel *The New Atlantis* of 1627. This woodcut by an unknown artist visualizes Bacon's utopia. Note, for example, the practically large strawberry or the small deer

taste (Fig. 3.4). Thus, as sedentism increased during the Neolithic Revolution, local farming, selection, and intervention in reproduction and the gene pool began. We can probably only speak of breeding in the modern sense from the eighteenth century onwards (Fig. 3.2).

Several findings play an important role here. In 1753, the Swedish botanist Carl von Linné presented a system for classifying all living things. Linné classified forms resulting from selection breeding, which we now call breeds or varieties, as subspecies. In parallel, in the second half of the eighteenth century, the British farmer Robert Bakewell, called *The Great Improver* in his days, made significant advances in the breeding of cattle, horses, and sheep. Furthermore, Charles Darwin published his selection theory *On the Origin of Species* in 1859, establishing the modern understanding of the processes and mechanisms of evolution [2]. He included the findings of Robert Bakewell in his theorizing and referred to him explicitly in the first chapter:

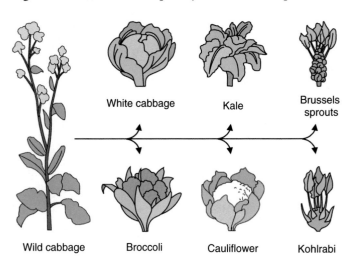

Fig. 3.2 Breeding of different cabbage varieties from wild cabbage *Brasica oleracea* by selection

Youatt gives an excellent illustration of the effects of a course of selection which may be considered as unconscious, in so far that the breeders could never have expected, or even wished, to produce the result which ensued—namely, the production of the distinct strains. The two flocks of Leicester sheep kept by Mr. Buckley and Mr. Burgess, as Mr. Youatt remarks, "Have been purely bred from the original stock of Mr. Bakewell for upwards of fifty years. There is not a suspicion existing in the mind of any one at all acquainted with the subject that the owner of either of them has deviated in any one instance from the pure blood of Mr. Bakewell's flock, and yet the difference between the sheep possessed by these two gentlemen is so great that they have the appearance of being quite different varieties."

If there exist savages so barbarous as never to think of the inherited character of the offspring of their domestic animals, yet any one animal particularly useful to them, for any special purpose, would be carefully preserved during famines and other accidents, to which savages are so liable, and such choice animals would thus generally leave more offspring than the inferior ones; so that in this case there would be a kind of unconscious selection going on. We see the value set on animals even by the barbarians of Tierra del Fuego, by their killing and devouring their old women, in times of dearth, as of less value than their dogs.

William Youatt, quoted by Darwin, was a founding member of the *Royal Agricultural Society of England,* founded in 1838, and published quite a hero in this field. Finally, the research results of the Moravian-Austrian abbot of St Thomas's Abbey, Brno Gregor Mendel revolutionized breeding. In 1866 he published his results on crossbreeding experiments in plants, especially the pea [3]. Right at the beginning of his treatise Mendel writes:

Experience of artificial fertilization, such as is effected with ornamental plants in order to obtain new variations in colour, has led to the experiments which will here be discussed. The striking regularity with which the same hybrid forms always reappeared whenever fertilization took place between the same species induced further experiments to be undertaken, the object of which was to follow up the developments of the hybrids in their progeny.

The regularity that was noticed eventually led to **Mendel's Laws of Inheritance**, which are still valid today and are taught in schools. In retrospect, it may seem surprising that Mendel's findings were not initially wider disseminated despite being published. It was not until 1900 that Mendel's observations were independently rediscovered and confirmed by three botanists, all of whom were involved in plant breeding, thus acknowledging the importance of his achievement. Mendel also knew Darwin's theory of evolution and incidentally refuted Darwin's view of the course of heredity. According to Darwin, traits inherited from parents continue to intermingle in subsequent generations, virtually thinning out. Mendel, on the other hand, shows that these traits remain constant and are passed on according to fixed rules. The picture of evolution—and breeding—in the twentieth century is ultimately based on the synthesis of Mendel's theory of heredity and Darwin's theory of evolution.

Mendel's demonstration that certain traits are regularly transmitted from a parent plant to its offspring was not only an important contribution to Darwin's theory of selection. It was a decisive contribution to being able to breed new varieties and breeds by selective choice of parent organisms. To speed up breeding, natural variation of traits soon became insufficient. Rapid progress was made, particularly in plant breeding, where there were few ethical concerns about using chemicals or radiation to introduce random genetic changes (mutagenesis).

3.1 Nuclear Gardening

As early as 1901, the Dutch botanist Hugo de Vries published his **mutation theory** (also known as mutationism or Mendelism), according to which new species arise suddenly, without transitions and in an undirected manner [4]. He mused that artificial mutations could be used to generate new, better-adapted and higher-yielding plants. On his own, he lacked the knowledge of how to trigger mutations. Although the German physicist and Nobel Prize winner Wilhelm Röntgen had already discovered the X-rays named after him in 1895, which could have served as a "tool" for de Vries, chance did not want to bring the two worlds together. The first observations that chemicals could trigger changes in the genetic makeup were made in 1910s. By the 1940s and 1950s, methods for chemically producing genetic changes (chemical **mutagenesis**) were widely established [5]. In parallel, methods were also developed to produce mutations using ultraviolet or ionizing **radiation** (X-rays and radiation released by radioactive decay) [6]. Famous and pioneering were the fruit fly mutants produced in this way by the US geneticist and Nobel laureate Hermann Muller in 1927 [7]. The aim of generating genetic variability as a result of chemical or radiation action was usually to study fundamental life processes. The comparison of intact and defective often provided information about function. Of course, the usefulness for breeding was also recognized (Fig. 3.3).

In 1928, the US geneticist Lewis Stadler was the first to induce mutations in barley seedlings using X-rays [8]. In 1934, the first radiation mutant came onto the market in the form of a **tobacco variety** from the Dutch colonies. The advantage of radiation over chemicals is its more far-reaching effect. It is not necessary to bring every single seed, every single plant into contact with the chemical—which is

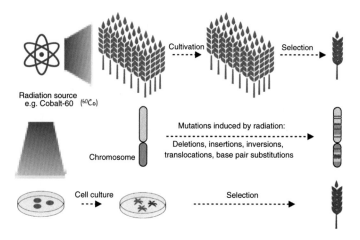

Fig. 3.3 Nuclear gardening. A radioactive source is used to create mutations in the chromosomes. This can be done on the intact plant or in plant cell cultures in the laboratory

also a risk for the experimenters. It is enough to place a radioactive emitter (usually a **cobalt-60 isotope,** which emits mostly gamma rays) on an experimental field and thus irradiate an entire plantation [9]. To protect employees, the cobalt sources were sunk into the ground under lead cover and only exposed by remote control when necessary. Such so-called **gamma gardens** were established as plant breeding laboratories in Europe, India, Japan, the USA and the USSR, for example. Breeding companies advertised their *atomic energized seed*s in newspapers and magazines.

In 1959, even an *Atomic Gardening Society* was founded in England by Muriel Howorth [10]. It was aimed specifically at the public in order to involve citizens in the raising and description of irradiated seeds (Sect. 7.1). Founded in Eastbourne, England, the Society operated worldwide with the aim of:

[...] atomic experimental gardens span the world, linking atomic plant mutation experimenters of many countries, working for the benefit of all with a benevolent effort to find more food for more people faster.

Hobby gardeners could also buy seed packets in garden shops, on which it was written that it is uncertain whether the seeds will sprout and what changes will be present: *This experiment may produce new types of plants. Perhaps you will find changes in size, colour, or shape.* In commercial use, mutagenesis triggered by atomic radiation brought more than 1800 plant varieties to market, according to an *International Atomic Energy Agency* list: in the food sector, mainly rice, wheat, oats, canola, corn, soybeans, peanuts, beans, olives, and many fruits and vegetables. Many of them are still cultivated today or have been cross-bred into new varieties.

One product of the gamma garden at New York's *Brookhaven National Laboratory* was the fungus-resistant **peppermint variety** Todd's Mitcham from the *A.M. Todd Company,* approved in the United States in 1970 [11, 12]. Nearly all globally traded peppermint oil for chewing gum, toothpaste, mouthwashes and confectionery comes from this variety. The genetic reason why Todd's Mitcham is fungus resistant is unknown. Interestingly, the source plant of the radiation was itself already a cultivar. The peppermint plant was first described in 1753 by Swedish naturalist Carl von Linné as *Mentha x piperita.* In the English district of Mitcham, southwest of London, peppermint has been cultivated since the mid-eighteenth century. However, the Mitcham variety was itself already naturally genetically modified: It is sterile and can only be propagated vegetatively from rootlets. The reason for this is that *Mentha × piperita* is a cross between water mint *(Mentha aquatica)* and spearmint (Mentha *spicata).* Whether this cross was created by chance or was bred too is not known.

Grapefruit varieties such as Star Ruby and Ruby Red are also products of gamma gardens. Almost all **barley** grown in Europe carries one of two genes that were modified by radiation decades ago to make the ears grow on shorter, sturdier stalks, making them more stable. This need not worry us, however. These varieties have been grown and marketed for more than half a century and there is no evidence that they have caused any damage to human or animal health. On the contrary, it is no exaggeration to say that without these varieties the world's food supply would be in jeopardy. And it does not matter whether we are talking about organic, integrative or conventional agriculture. Genetically speaking, mutagenic treatment leads to an unimaginable number of changes in the genetic material. Sections can be lost, multiplied, rearranged or altered at any point in their sequence (Fig. 3.4). This does not matter for market approval. Europe's agencies do not look at the genotype, but only the **phenotype**, the appearance and properties. Legally, mutagenesis with chemicals or radiation is exempt from the German Genetic Engineering Act (Gentechnikgesetz). As with conventional breeding, the plants are subject to a correspondingly simplified approval procedure.

3.2   Conventional Breeding Methods

With *atomic gardening*, we have become familiar with a method of conventional breeding (Fig. 3.4), in this case mutation breeding. Although plants produced by **mutation breeding** are classified as genetically modified organisms (GMOs) in the European Union, they have a **special position** in that they are exempt from all authorisation and labelling requirements applicable to GMOs. According to the **ruling of the European Court of Justice** (ECJ) of July

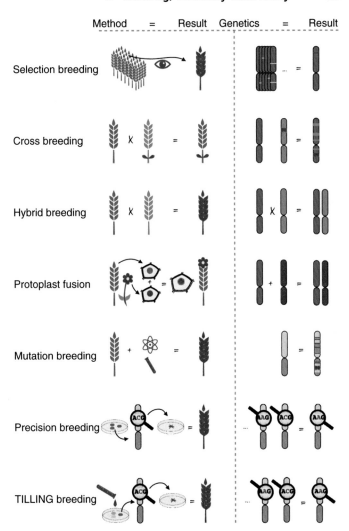

Fig. 3.4 Conventional breeding methods. All these methods are by definition not classified as genetic engineering yielding GMOs. Explanations can be found in the text

25, 2018, however, this exemption only applies to the described undirected mutation breeding but not to the new but far more precise gene editing with the CRISPR/Cas system (Sect. 5.1) [13]. Undirected mutation breeding gives rise to new varieties, the commercial cultivation and marketing of which requires **variety approval** by the Federal Office of Plant Varieties in Germany. Prerequisite for the approval of a new variety is its distinctness (D) from other varieties of the same species as well as its uniformity (U) and stability (S). This is tested by cultivation in the open field or in the greenhouse (**DUS testing**). In the case of the most important agricultural plant species, it is also necessary to check their value (V) for cultivation (C) and use (U) (**VCU testing**). This is given if, in the totality of its value-determining characteristics, a clear improvement for plant cultivation, for the utilisation of the harvested crop or for the utilisation of the products obtained from the harvested crop can be expected in comparison with the approved comparable varieties.

Conventional breeding methods in the narrower sense are thus based on the selection of partners to be crossed and a selection of offspring on the basis of certain measurable characteristics (Fig. 3.4). This also includes so-called **hybrid breeding**, in which a targeted crossing of two parental lines is carried out. In the parent lines, the desired traits must be present in as homogeneous as possible (inbred lines). Due to the largely not understood so-called **heterosis effect**, the offspring are often more productive and resistant than their parent lines. In the following generation, however, the effect is lost and the farmer has to buy new seed.

A further development of classical crossbreeding is **precision breeding** (also known as *SMART breeding,* where SMART stands for *selection with markers and advanced reproductive technologies,* or **marker-assisted breeding**). In this method, the partners to be crossed or the plant to be chosen are selected on the basis of a genetic analysis *(selection with markers)* (Fig. 3.4). This assumes that the breeder already knows which genotype he is looking for and how it relates to the desired phenotype. The main advantage is that the plant does not have to grow to maturity, but the seedling can already be examined. Using methods of cell culture technology *(advanced reproductive technologies),* a complete plant can then be regenerated from individual cells.

Cell culture techniques are also used in **protoplast fusion** (also called **somatic hybridization**) (Fig. 3.4). This is a conventional breeding method in the broader sense, as it goes far beyond the methodology of classical crossing. The protoplast is the name given to a plant cell stripped of its cell wall. Unlike animal cells, which are rather wobbly due to their flexible cell wall, plant cells form solid cell walls are interspersed with cellulose. After their enzymatic degradation in the test tube, the protoplast remains. It is surrounded only by a cell membrane. Two plant cell protoplasts can now be fused together in the test tube. Under the right conditions, the cell nuclei then also fuse together with the genetic information they contain and can be regenerated into complete plants. This form of cell crossing is also possible across genus boundaries. The result can be, for example, a *pomato* (a cross between potato and tomato with no agronomic benefit) or a virus-resistant asparagus variety. Somatic hybridization is also possible with human and animal cells. For example, human and murine, that is human and mouse cells, were successfully fused together for experimental purposes as early as the late 1960s [14]. The fusion cells proliferate in appropriate culture media but do not grow into a differentiated organism. The method was previously used to assign genes to specific chromosomes.

Another highly modern, but in a broader sense conventional, non-GM breeding method is the **TILLING method** (*targeting induced local lesions in genomes*), which has been established since the 2000s. This method also assumes that genetic information of the plant is already available and that it is known in advance which genetic change is to be achieved (Fig. 3.4). As in undirected mutation breeding, individual nucleotides are exchanged by means of a chemical (usually ethyl methanesulfonate; EMS for short)—so-called point mutations are produced. This is followed by a clever *screening* procedure, which we already know from precision breeding and which is based on reading the DNA (Chap. 4). In this way, seedlings can be identified that contain desired changes at desired locations in the genome. As a result, plants can be bred that do not differ from those developed using modern methods such as gene editing with the CRISPR/Cas system. However, politicians and society are usually only interested in whether a newly bred organism is a genetically modified organism in the sense of the legal definitions or not. Approval procedures and consumer decisions largely depend on this classification (Sect. 3.7).

3.3   Colourful Genetic Engineering

As we have seen, modern conventional breeding also makes use of genetic engineering knowledge. Before we turn to the application of genetic engineering in plant (Sect. 3.4) and animal breeding (Sect. 3.5), I would like to briefly remind you that bacteria are also bred, for example for biotechnological applications. Depending on the area of application, genetic engineering has been given colourful names. It is obvious that genetic engineering for plants is called **green genetic engineering**. Likewise, the designation **red genetic engineering** for applications in medicine and **blue genetic engineering** for applications with sea creatures. More imagination is needed for **white genetic engineering** for industrial applications, **brown** for waste management applications and **grey** for ecological applications. Of course, there are also genetic engineering applications in animals, but no colour has yet been assigned to them. This may also be due to the fact that the areas of application often overlap. I therefore take the liberty of using the term **mottled genetic engineering** for their application in animals. We will then turn to the application in humans from Chap. 5 onwards.

At this point, I would like to briefly provide some general observations on genetic engineering and its application in everyday products. It is impossible to imagine our everyday life without genetic engineering. There are very obvious applications, such as when a food package states that it contains oil from genetically modified soy plants. Strict rules apply to the food industry in the European Union with regard to **labelling requirements**. The general rule is that if a food ...

- ... is a genetically modified organism (**GMO**), it must be labelled. Meat from genetically modified cattle or corn would therefore have to be labelled.
- ... **contains GMOs.** Yoghurt with genetically modified bacteria or beer with genetically modified yeast are subject to labelling.
- ... is **produced from GMOs.** Edible oil from genetically modified soya or rapeseed plants, sugar from genetically modified sugar beet or starch from genetically modified maize must be labelled.

There are also **exceptions** to labelling, namely when ...

- ... the proportion of genetically modified ingredients does not exceed **0.9%** and the manufacturer can prove that this proportion is adventitious or technically unavoidable. This is interesting in the case of honey, as the only ingredient in honey that requires labelling is pollen, as this is the only ingredient that contains DNA. However, honey rarely contains more than 0.5% pollen. Therefore, even if all pollen came from transgenic plants, honey would not be subject to labelling.

- ... the food or ingredient was **produced with the aid of GMOs.** This also includes meat, milk or eggs from animals that have received feed from genetically modified plants or animals. The feed itself, however, is subject to labelling.
- ... additives such as flavours, vitamins, citric acid or the flavour enhancer glutamate, which were produced with the help of GMOs but no longer contain GMOs.
- ... technical additives or enzymes from GMOs were used for the production of the food. For example, the enzyme chymosin is needed in cheese production. Traditionally, it is obtained from calves' stomachs, but nowadays it is mainly obtained with the help of genetically modified microorganisms.

Experts in the food industry estimate that applications of genetic engineering not subject to labelling play a role in around 70% of the foodstuffs on the market. In addition to adventitious and technically unavoidable traces of authorised GMOs, these are mainly additives, vitamins, amino acids, enzymes and other auxiliary substances produced with the aid of genetically modified microorganisms. In fruit and berry processing for **juice and wine production,** for example, enzymes such as amylase, arabinase, glucoamylase, pectinases, pectin esterase, pectin lyase, polygalacturonase, proteases and rhamnogalacturonase are used. Those who press their juice at home do not need these enzymes. But for large-scale production, these enzymes are necessary, among other things, to achieve a higher juice and aroma yield during pressing, to support clarification after pressing or to increase filterability. The quantities used are in the gram range per hundreds liter of product.

However, enzymes do not only play a major role in food production. Other important areas of application are the **paper**, **textile**, **cleaning and cosmetics industries**. **Toothpaste** contains enzymes that help clean teeth. In contact lens care products, enzymes break down fat and protein deposits. **Detergents** contain enzymes to remove proteins and starch. The list goes on. The majority of enzymes are produced with the help of genetically modified microorganisms. Why? Often the enzymes used come from specific tissues or organs of animals or plants, which often makes them difficult to extract. Instead, the enzyme-coding genes are transferred to a bacterium or fungus that can be propagated relatively easily and in large quantities. Often, the organisms are designed to deliver the product into the culture medium, making it fairly easy to purify and sell. In addition, the enzymes are modified in their structure, usually by altering the genetic code, so that they are **optimized** for their task. At warm to hot temperatures, enzymes, which are proteins, would denature like a fried egg in a hot pan, that is they irreversibly change their structure. A wash cycle at 60 degrees Celsius would then be the end for the enzyme. However, genetically optimized enzymes can withstand these temperatures. Conversely, enzymes have been developed that also work effectively at low temperatures. This means that the temperature can be reduced during the wash cycle and electricity can be saved.

In the **textile industry**, it is the raw material **cotton** as well as additives for dyeing and fabric processing (lubricants) that are often obtained from genetically modified organisms. But cotton does not only end up in textiles. The value chain goes much further. After harvesting, the fibres are separated from the protein- and fat-rich seeds. This produces by-products that are used as food and feed. For example, the high-quality oil from the cotton seeds is used as cooking oil for deep-frying or in the production of margarine. The protein-rich meal is used as feed and can be used as a basic material for obtaining cotton milk for the cosmetics industry. Various food additives such as cellulose (E460) or methyl cellulose (E461) are obtained from the very short, non-spinnable, cellulose-rich cotton fibres. They serve the food industry as thickeners, stabilizers, emulsifiers or fillers. However, the main customer for the fibres is the paper industry. They are mainly used to produce high-quality, tear-resistant paper, for example for banknotes.

The **paper industry** also has a great need for high quality **starch** to produce, for example, glossy or photographic paper. Starch is a long-chain molecule that is very viscous in aqueous solution. That is why we also bind sauces with starch. However, a starch solution has another special property: it behaves as a non-Newtonian fluid, which means that it has a variable viscosity. In extreme cases, high and jerky local pressure causes a starch solution to behave like a solid. This would clog nozzles used to spray the starch solution onto the paper substrate during the production process. Therefore, the solution is treated (conditioned) with enzymes. And these are in turn produced with the help of transgenic microorganisms.

If one takes the trouble to follow the processes of creation of consumer goods in an average European household, it becomes clear how far genetic engineering has already become a matter of fact.

3.4   Green Genetic Engineering

Those who speak against genetic engineering usually have **green genetic engineering** in mind, that is the application of genetic engineering in agriculture (Figs. 3.6 and 3.7). It is true that Germany and large parts of Europe are *genetic engineering-free zones*. However, this only refers to the cultivation of genetically modified plants, not to their import or the application of genetic engineering in industry. The first plant approved for commercial cultivation and sale was a virus-resistant **tobacco** in the People's Republic of China. The first plant approved and cultivated for consumption was the **FlavrSavr tomato** (flavour-preserving tomato), which was cultivated and marketed in the USA from 1994 to 1997. However, it was rejected by the consumers.

No genetically modified plants are currently cultivated in Germany. The last plants approved for commercial cultivation were *BASF*'s **Amflora potato** for the starch industry (until 2011) and *Monsanto*'s insect-resistant **Bt maize MON810** (until 2008). In 2012, genetically modified sugar beets and potatoes were still grown for research purposes, and in 2013 the genetically modified **bacterium** *Rhodococcus equi* strain RG2837 [16]. The latter, of course, was not grown: 120 foals were inoculated with the bacterium to test a novel way of combating pneumonia in horses. Nevertheless, products of genetically modified plants are **imported** into the European Union and Germany on a large scale, first and foremost: soya. Europe produces too few protein-rich **forage** for its livestock and is therefore dependent on the import of large quantities of soya beans. In purely mathematical terms, most of these beans can only come from genetically modified **soybean**, as the amount of non-transgenic soya plants cultivated in the European Union is much lower than the demand. Worldwide, the proportion of genetically modified soybean production is just under 80%. Currently, 19 different transgenic soybean varieties are approved for **import** into the European Union. Also approved for import into the European Union as of March 16, 2019 are 52 varieties of corn, 13 varieties of canola, 13 varieties of cotton, seven varieties of ornamental flowers (garden carnations) and one variety of sugar beet. This data shows how dependent we have become on transgenic crops. But what are the basics of genetic engineering?

The origins of genetic engineering date back to the early 1970s. The US scientist Stanley Cohen from Stanford researched the mechanisms that lead to antibiotic resistance in bacteria. He discovered that bacteria carry extrachromosomal genetic information for resistance, in addition to the genetic information on the chromosomes. This additional genetic information is located on DNA molecules that have been known since the 1950s and are called **plasmids** [17]. These can be easily exchanged among bacteria, which is why antibiotic resistance, for example, can spread rapidly. Cohen succeeded in carrying out this exchange in the laboratory. He developed a method that allowed him to isolate plasmids from bacteria and introduce them into other bacteria. In parallel to Cohen's work, the US scientist Herbert Boyer from San Francisco was researching molecules with which bacteria protect themselves from viruses. These are so-called **restriction enzymes** with which DNA is cut at very specific nucleotide sequences. This results in single-stranded overhangs with a specific nucleotide sequence at the ends of the double-stranded DNA. This sequence in turn depends on the restriction enzyme used. During a conference on the island of Hawaii in 1972, at which both scientists presented their research work independently of each other, something clicked for both of them: If a plasmid is opened with a restriction enzyme and DNA from another organism is treated with exactly the same restriction enzyme, then both fragments can form base pairs at the overhangs and fuse (**ligation**). This again creates a closed DNA ring, a hybrid of the bacterial plasmid and the organism's DNA fragment. This is **recombinant DNA** (rDNA). The hybrid plasmid can be reintroduced into bacteria (**transformation**) and propagated there (**cloning**) using the method developed by Cohen. In 1974, the first transgenic organism was created in this way [18]. This was an *Escherichia coli* bacterium that contained a fragment of the genome of *Staphylococcus aureus* in a plasmid of the bacterium *Salmonella panama* (the plasmid is named pSC101, with p for plasmid and SC for Stanley Cohen). On November 4, 1974, Boyer and Cohen filed an application for the process with the U.S. Patent Office. In 1980, they were granted the first **patent**, and others followed [19]. This meant that all scientists who inserted any DNA fragment from any organism into any plasmid had to pay royalties. This earned Stanford and San Francisco Universities, and proportionately Boyer and Cohen, about US$300 million over the life of the patent [20]. This example also motivated other universities around the world to encourage their researchers to patent processes—until today. In 1976, Herbert Boyer and investor Robert Swanson founded the first genetic engineering company, ***Genentech***, which still exists today and launched the first recombinant **insulin** in 1978. Prior to that, insulin was derived from the pancreases of cattle or pigs, which often led to immune response in patients. It is worth mentioning that the scientists were aware of the novel process they had developed. For the most part, they were also aware of their responsibility (Sect. 3.7).

A natural process was used to produce the first genetically modified plants. At the end of the 1970s, the Belgian molecular biologists Jeff Schell, Director at the *Max Planck Institute for Plant Breeding Research* in Cologne from 1978 to 2000, and Marc van Montagu described a mechanism by which the soil bacterium ***Agrobacterium** tumefaciens* infects plants, thereby incorporating bacterial DNA into the plant genome (Fig. 3.5) [21].

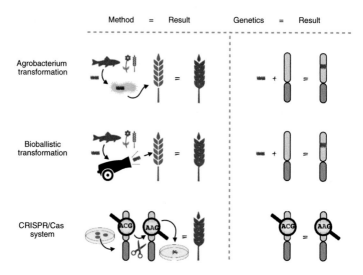

Fig. 3.5 Genetically assisted breeding methods. All the methods shown are classified as genetic engineering by definition, thus yielding GMOs. All methods are suitable for cis- and trans-genetic modifications. The CRISPR/Cas system is shown in more detail in Fig. 5.2. Explanations can be found in the text

The genetic information that the bacterium uses to genetically transform the plant is located on a plasmid, the so-called **Ti plasmid** (*tumour inducing*). It is about 200,000 base pairs long and carries all the information necessary to infect a plant and cause it to develop tumours, which the bacteria then colonise. Through the fundamental work of Boyer and Cohen, recombinant work with plasmids was well known and the first transgenic plants were created in the 1980s [22]. In 1986, the first with genetically modified insect-, bacteria- and virus-resistant plants took place in the USA and France. However, before a genetically modified plant can be tested in the field, intensive laboratory and greenhouse tests are required. These often take years (Sect. 7.2). It is only under field conditions that the interaction of the genetic information newly introduced into the plant with the complex conditions of nature is finally proven. Between 1986 and 1997, some 25,000 field trials were carried out worldwide with over 60 different plants in 45 countries. The People's Republic of China approved the first **commercial cultivation** of a transgenic, virus-resistant **tobacco plant** in 1992. This was followed in 1994 by the first fruit approved for consumption in the USA (the **FlavrSavr tomato**, colloquially known as the anti-mud tomato) and in the European Union by the approval of a tobacco plant resistant to an herbicide. Since then, the areas cultivated with genetically modified crops have steadily increased (Figs. 3.6 and 3.7) and numerous other methods have been developed to integrate foreign DNA into a plant genome [15].

Most of the processes are based on natural mechanisms, which are modified accordingly for technical use. One exception is the **gene gun**. Here, DNA is applied to gold or

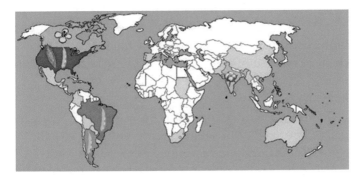

Fig. 3.6 Countries in which genetically modified plants are currently cultivated. Of the total of almost 190 million hectares worldwide, 40% are in the USA (mainly corn and soybean), 26% in Brazil (mainly soybean), 12% in Argentina (mainly soybean), 7% in Canada (mainly rapeseed) and 6% in India (mainly cotton). Areas in Spain and Portugal account for less than 1% and are limited to the cultivation of the Bt maize MON810. (Source: ISAAA [15])

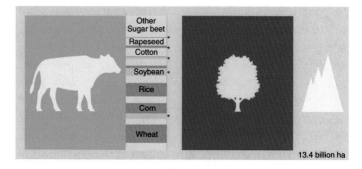

Other
Sugar beet
Rapeseed *
Cotton
·
Soybean ·
Rice
·
Corn
·
Wheat

13.4 billion ha

Fig. 3.7 The total area of the Earth is about 13.5 billion hectares (135 million square kilometres). Of this, 3.9 billion hectares are forest and five billion hectares are agricultural land. Of this, about 71% is pasture. Of the 1.45 billion hectares of arable land, again 71% is used for forage and about 18% for food crops (260 million hectares). Genetically modified crops are grown on about 13% of the arable land. The most common feed and food crops are highlighted on arable land. Percentages of land with genetically modified crops are shown lighter and highlighted with an asterisk. (Source: ISAAA [15])

tungsten particles and then shot into the cells at speeds of more than three times the speed of sound. The aim of this **bioballistic DNA transfer** (Fig. 3.5), which is a purely mechanical method of gene transfer, is the subsequent, again natural, incorporation of the DNA by the plant into its genome.

The only plant developed in Europe that has been partially approved for cultivation (there is no approved animal) is *BASF*'s **Amflora potato**. Among other things, a gene was switched off so that only one form of starch, preferred by the starch industry, is produced: amylopectin. Amylopectin is mainly used in the cement, fabric and paper industries as a lubricant or coating agent. For industrial use, the two types of starch, amylose and amylopectin, had to be separated from each other by chemical, physical or enzymatic processes. The approval process took 14 years, from August

1996 to March 2010. In 2008, scientists at the *Fraunhofer Institute for Molecular Biology and Applied Ecology* created a potato with the same properties using the TILLING process described above (Sect. 3.2 and Fig. 3.4), which does not require approval [23]. How is this to be assessed? First of all, it is quite simple: genetic engineering methods were applied to Amflora, in particular, DNA segments foreign to potatoes were introduced into the genome. This makes it clearly transgenic and it must be described genetically precisely for the approval procedure and approved for cultivation. The *Fraunhofer* scientists subjected potatoes to chemical mutagenesis as described, examined young shoots genetically with regard to starch metabolism, selected suitable mutants and raised them. Thus, genetic characterization was not necessary and only variety approval procedures were required (DUS and VCU testing).

Using CRISPR/Cas-based gene editing (Fig. 3.5), scientists from Sweden and Argentina developed a potato in 2018 that also only produces amylopectin [24]. The four affected gene segments are missing between one and twelve base pairs (DNA building blocks). No foreign DNA was introduced into the potato genome. Prior to the ECJ ruling in July 2018, this potato would not legally be a genetically modified organism and would also have only required a variety authorisation for commercial exploitation. Some European countries, such as Sweden, had also decided in this way and in 2015 released already gene-edited plants for experimental cultivation without an approval procedure [25]. This is now no longer permitted.

In genetic engineering, a distinction is often made between cis genetic engineering and trans genetic engineering. The Latin prefix *cis* stands for "on this side" and the prefix *trans* for "on the other side". In **cis genetic engineering**, no DNA fragments or genes from other species are used. The genetic modification is therefore limited to the target organism. This is the case, for example, when a gene is deactivated, modified or incorporated in multiple copies in an organism. In **trans genetic engineering**, on the other hand, DNA fragments from any other species are introduced into the target organism. This is the most common form of genetic engineering. CRISPR/Cas can be used for both cis- and trans-genic work (Sect. 5.1 and Fig. 5.2). Increasingly, transgenic plant varieties are also being approved that contain several new traits. This means that several traits are stacked. For example, the maize variety MON87419 is approved for cultivation in the USA and Canada and in numerous other countries at least for use as feed or for human consumption. Two different herbicide resistances to weed control agents with the active ingredients glufosinate or dicamba were introduced into this variety.

As we have seen, conventional breeding methods in the broader sense (Fig. 3.4), which are not subject to the Genetic Engineering Act and thus to genetic analysis, include numerous methods that intervene deeply in the organisation of the genome. The German law regulating genetic engineering, the **Genetic Engineering Act**, dates from 1990 and regulates, as described in Sect. 3.7, the use as well as the prevention of hazards. The ordinance on safety levels and safety measures for genetic engineering work in genetic engineering facilities *(Genetic Engineering Safety Ordinance—GenTSV)* specifies how this Act is to be implemented in practice. It was last amended in 2015, although the changes made were only of an organisational nature and not in terms of content. Therefore, nothing has changed with regard to the assessment criteria.

According to the German Genetic Engineering Act

a genetically modified organism [GMO] means an organism, other than a human being, whose genetic material has been altered in a way that does not occur naturally by means of cross-breeding or natural recombination; a genetically modified organism is also an organism resulting from cross-breeding or natural recombination between genetically modified organisms or with one or more genetically modified organisms, or from other means of propagation of a genetically modified organism, provided that the genetic material of the organism has characteristics resulting from genetic engineering,

and it explicitly excludes

[...] (a) mutagenesis; and (b) cell fusion (including protoplast fusion) of plant cells from organisms capable of exchanging genetic material by means of conventional breeding techniques [...].

The main problem of the current version of the law lies in the words *does not occur naturally by means of cross-breeding or natural recombination* . Single nucleotide exchanges, so-called **point mutations**, do occur naturally and can—as described in Sect. 3.2—be caused in an undirected manner by ionising radiation or chemicals. Therefore, mutagenesis is excluded. Gene editing using the CRISPR/Cas system (Sect. 5.1) is a mutagenesis procedure that is, however, classified as genetic engineering (Fig. 3.5). If scientists are presented with two organisms in order to decide which has been created by classical mutagenesis and which by gene editing, they will fail. This has been the basis in many non-European countries—and was the basis in some European countries before the ECJ ruling on July 25, 2018—for not regulating CRISPERed organisms. In the USA and Canada, the entire authorisation system is based on this **equivalence principle**, or a **product-oriented** authorisation procedure: if a product does not differ from an already authorised or natural product, it is authorised or not regulated. In all other countries where there are statutory approval procedures for new plant or animal varieties, a **procedure-oriented** approval procedure applies (Fig. 3.8).

For example, at the end of 2018, the first gene-edited **soybean variety** was approved for marketing by the company *Calyxt* on an initial 6700 hectares in the USA. Using CRISPR/Cas gene editing, two targeted mutations were introduced, changing the fatty acid quality in the beans. They now contain less saturated fatty acids and more of the monounsaturated oleic acid. As a result, the soybean oil extracted from these beans forms fewer trans fatty acids when heated, such as during frying or deep-frying. These in turn are suspected of increasing the so-called LDL cholesterol level (*low density lipoprotein*) in the blood and thus promoting heart attacks and arteriosclerosis. For this reason, foods containing trans fatty acids are subject to mandatory labelling in the USA. Products derived from the gene-edited soybean are exempt from this labelling. They also do not have to be labelled as genetically engineered because, as just described, they do not fall under the Genetic Engineering Act according to the equivalence principle.

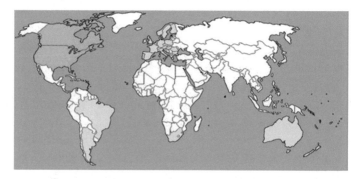

Fig. 3.8 States with clearly defined process-oriented (yellow) or product-oriented (green) approval procedures for the marketing of new varieties. Various approval procedures apply in the other countries. (Source: Eckerstorfer et al. [26])

In Europe, the situation is different. The ECJ justifies the legal distinction between new and old mutagenesis methods by stating that the new methods enable *the production of genetically modified varieties at a much greater rate and to a much greater extent* than is possible with radiation or chemicals. Since mutagenesis by means of the CRISPR/Cas system cannot be detected in the product, it will hardly be possible to implement the ECJ's ruling [27]. How is that supposed to work? Company controls? How are the import of such organisms or products such as soybean oil to be controlled? That is impossible so far. There is an international so-called Biosafety Clearing-House. It is a kind of registry for genetically modified organisms and information on the genes used as well as laws and other relevant data. Participation in the Clearing House process, although stipulated in the 2003 International Biosafety Protocol, or **Cartagena Protocol**, is still voluntary. Moreover, the United States, as the world's major producer of new breeds, has not signed the protocol. Alternatively, one could think about a kind of "Green Helmets", analogous to the peacekeeping forces of the United Nations—without weapons, of course, but with access to confidential company documents. But here, too, the legal situation would be foreseeably difficult. For these reasons, organisms and derived products imported into Germany would have to be strictly labelled *may contain genetic engineering.*

The **labelling** of genetically modified food is worth some discussion anyway. What is the aim? The most important thing is surely transparency. The consumer should be given the opportunity to choose. However, this freedom of choice is only based on the definitions in the Genetic Engineering Act. And this, as we have seen, excludes other interventions in genetic material. This is framing on the part of politics. It would be desirable to extend transparency to other breeding methods that actively interfere with genetic material (Sect. 3.7).

Particularly in plant breeding, the area of application of genetic engineering most intensively perceived by consumers, it is important to take into account another important point influencing public perception: the fact that at present essentially only herbicide and insect resistance is used as a produced trait on agricultural land worldwide.

Herbicide-resistant plants are largely insensitive to weed control agents (herbicides). This makes it easier for farmers to control weeds. Instead of spraying with pinpoint accuracy, the farmland can now be treated over a large area. Of course, this also makes it easier for chemical-related seed companies to market the crop plus herbicide combo package. Combined with the major hurdles in the approval process, this opens the door to monopolies. Currently, the most widespread herbicide resistance is to the active ingredients **glyphosate** (RoundupReady from *Bayer,* formerly *Monsanto*) and **glufosinate** (LibertyLink from *BASF*), which has been bred using genetic engineering methods. Resistance is usually achieved by introducing one or more genes that break down the herbicide in the plant and thus render it harmless. Conversely, genes and thus the gene products that form a target for the herbicide can be inactivated. Of course, this assumes that the target protein does not have an indispensable function for the plant. Therefore, this is the rarer method.

Insect-resistant plants produce their own insecticide, that is a substance that kills insects and possibly also larval stages of these insects. The only insecticidal substance of note used in genetically engineered plants is Bt toxin, also known as delta endotoxins (*cry*-toxin*).* The **Bt toxin** is a protein originally discovered in bacteria of the species *Bacillus thuringiensis. The* designation Bt is derived from the name of the bacterium, the designation *cry* from the fact that the protein occurs in the bacteria (or more precisely in their spores) as a *crystal.*

Using bacteria to combat insects is not new [28]. Even the pharaohs of Egypt are said to have used bacterial solutions, probably rather unconsciously, against insect plagues [29]. The first known systematic studies on the use of *Bacillus thuringiensis* in Europe date back to the 1960s [30]. In organic farming, according to the guidelines of the *Association for Organic Agriculture,* both the bacteria and Bt preparations are approved as plant protection products [31]. Transgenic Bt plants have been cultivated since 1996, initially mainly maize, cotton and potatoes. At that time, the main focus was on the control of the oak moth *(Tortrix viridana)* in Hessian forests. There is not only one Bt toxin—in fact, more than 89 different genes have already been discovered in different *Bacillus thuringiensis* strains, which produce just as many different toxin proteins. Some have a highly specific effect against a particular species or family of insects, while others have a non-specific effect as a broad-spectrum insecticide. The toxin exerts its effect in the gut of the butterflies, moths, flies, mosquitoes, wasps, bees and beetles that ingest it. This can happen naturally when food covered with bacteria is ingested, or when a Bt plant is eaten. It is then activated in the gut and destroys the gut walls.

In both techniques, the herbicide-resistant and the insect-resistant genetically modified plants, a toxin is at the centre. In the first case, a weedkiller; in the second, the Bt toxin. And poisons generate emotions—all the more so when it comes to their use in the environment and in food. And emotions can lead to irrational or selective evaluations (Sect. 3.7). Imagine a transgenic carnivorous arable plant that specifically attracts bark beetles or mosquitoes with a scent (pheromone) and releases the nitrogen-rich food into the soil as fertilizer. Or think of the many heat and drought tolerant plant varieties currently in development that have nothing to do with poisons. I am sure many of you would at least come to a different assessment of need. But these are hypothetical considerations. After all, we hardly ever talk about the concerns that farmers have elsewhere in the world. In the 1990s, for example, almost the entire **papaya population** on the Hawaiian island of Puna was threatened by the *papaya ringspot virus* [32]. As a result, the papaya crop had collapsed by more than half. A transgenic, virus-resistant variety called Rainbow has brought papaya farmers rising crop yields again and is still seen today as a success story of agricultural genetic engineering.

A similar problem is currently happening to the **banana** [33]. It is not just a tasty snack for us, but a staple food for around half a billion people in tropical and subtropical latitudes. In Germany, it is the second most popular fruit, after apples—the average German eats 10 kilograms per year. As early as the middle of the twentieth century, the banana harvest was decimated by the **Panama disease** (also called tropical *race*) caused by a fungus. Within a decade, 99% of the world's crop was destroyed. This was countered by the search for resistant banana varieties. In the 1960s, plantations switched from the Gros Michel variety, which had been common until then, to the more resistant Cavendish. Around 99% of exported bananas are currently of the Cavendish variety. Of course, there are more varieties, about 1000 are known, of which about 300 are edible. Today, however, practically all traded bananas are descendants (clones) of a single banana tree. This was cultivated by the sixth Duke of Devonshire William Cavendish in a greenhouse in England at the beginning of the nineteenth century. Now even this originally resistant variety is endangered. A new fungus is currently spreading in Africa and Asia and there is no suitable spraying agent (fungicide). Since Cavendish is propagated vegetatively, that is by cuttings, resistance cannot develop naturally. The necessary genetic variation is lacking (Fig. 2.6). In 2017, Australian scientists succeeded in genetically engineering a resistance gene from a wild banana variety into the Cavendish variety [34]. This makes Cavendish genetically immunized, so to speak. Interestingly, the resistance gene is already originally present in the Cavendish variety, but with ten times less activity. At present, attempts are being made to increase the activity of the existing resistance gene using CRISPR/Cas gene editing, instead of using genes from wild banana varieties, as has been done in the past.

3.5   Mottled Genetic Engineering

Animals are closer to us than plants and, accordingly, attitudes towards the use of genetic engineering in animals are different and approvals are manageable—at least when the products are aimed at humans as medicines or food. According to a 2018 poll in the U.S., the majority of the public has no concerns with the use of genetic engineering in animals when it benefits human health [35]. Release of transgenic mosquitoes, for example, which no longer transmit the malaria pathogen (Sect. 6.2), meets with broad approval. However, a corresponding field trial in Florida, on which the residents had to vote in parallel with the presidential election in November 2016, did not meet with approval. This is a typical picture that not only concerns the acceptance of genetic engineering: everything is acceptable as long as I benefit but am not directly affected. This is probably also the explanation for the broad approval given to the cultivation of human organs in animals that could then be transplanted into humans (**xenotransplantation**). In March 2019, the Japanese government approved a research program to this effect, in which hybrid pig embryos will be bred and brought to birth, developing organs from human tissue [36]. The aim is to fill the shortage of donor organs in this way.

Currently, there is no approved animal food in Germany. In Canada and the USA, only one transgenic **salmon** has been approved for consumption since 2015, in Canada without mandatory labelling. However, it has not yet found its way onto the European market. This Atlantic salmon, developed by the company *AquaBounty Technologies* back in 1989, may only be bred in closed facilities. It contains additional genetic information for two growth hormones from Chinook salmon and a gene regulator from the ocean pout. As a result of these changes, the fish grows much faster. It reaches more than twice the weight of unmodified Atlantic salmon (wild types) after just one year and its slaughter weight of around six kilograms after about 20 months. The approval process has taken 25 years [37]. It has been held up time and again, mainly because there was concern about public acceptance.

The situation is quite different with **GloFish** from the US company *Yorktown Technologies* (Fig. 3.9). After only two years of development, it was the first transgenic organism to be launched on the US market in 2003. GloFish is a transgenic zebrafish to which genes for the formation of a fluorescent protein have been added. This protein is excited to glow by ultraviolet light, known as black light. Some know this from disco, where lipsticks with fluorescent dyes are used, for example. So, there is also from the GloFish a whole range of different varieties for the home aquarium to buy.

In addition, two drugs derived from transgenic goats and chickens are approved in the USA. The human antithrombin III, an inhibitor of blood clotting approved in February 2009, is produced by goats in their **milk**. The drug in question is called ATryn, is marketed by *GTC Biotherapeutics* and is aimed at US citizens with a rare blood disorder.

Fig. 3.9 This ornamental fish was the first genetically modified animal to be approved in the USA in 2003. (Photo: www.Glo-Fish.com)

About one in 5000 citizens is affected. The other drug, Kanuma from *Alexion Pharmaceuticals*, is made by chickens in **chicken egg white** (approved, also in Europe, since December 2015). This drug is also aimed at patients with a rare metabolic disease.

The range of methods by which genes can be introduced into animals is diverse. An overview is shown in Fig. 3.10 [38]. In **pronuclear injection**, the genetic information to be transferred (cargo gene) is injected into the fertilised egg cell [39]. It can then be incorporated into the genome (transfection). Optionally, the components of the CRISPR/Cas system (Sect. 5.1) can also be injected at this time. **Sperm-mediated gene transfer** (SMGT) is a gene therapy procedure in which the cargo gene is injected into a sperm [40]. The loaded sperm is then injected into an egg by **intracytoplasmic sperm injection** (ICSI, Fig. 4.13). This allows the cargo

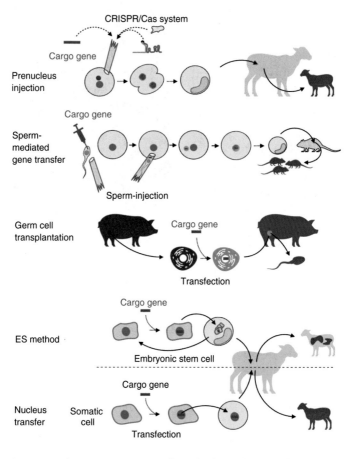

Fig. 3.10 The common range of methods used to produce genetically modified animals. The animals are freely selected here

gene to integrate into the genetic material. In **germ cell transplantation** (GCT), spermatogenic germ cells are removed from the testis, transfected in the laboratory and implanted into an infertile animal (Fig. 3.11) [41]. In the **ES method**, an embryonic stem cell is removed from a blastocyst, loaded (transfected) with the cargo gene, and injected back into a blastocyst. After delivery, a somatic mosaic animal is created, with modified and unmodified cells. In **nucleus transfer**, a somatic cell (somatic cell) is transfected and its nucleus is inserted into an enucleated egg cell (Fig. 3.12).

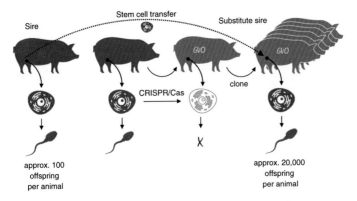

Fig. 3.11 Use of CRISPR/Cas system in animal breeding, illustrated by the pig. Sires (blue) are sterilized by knocking out a gene (green). These gene-edited animals can be cloned in the early developmental stage shortly after fertilization (blastocyst stage) (production of identical twins). These are implanted with sperm-producing germline stem cells from the desired sire (red) (germ cell transplantation)

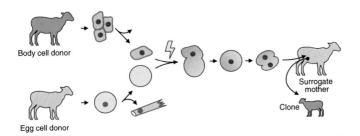

Fig. 3.12 In animal cloning, genetically identical copies are created. For this purpose, the cell nucleus with the genetic material is removed from a donor egg cell. A body cell of the animal to be cloned is fused with the nucleated egg cell by means of an electric shock and stimulated to divide. The cell cluster is implanted into a surrogate mother, who then gives birth to the clone

There are also numerous other traits that have been introduced into livestock using genetic engineering methods but never reached the market. These include enhanced **growth**, altered **fatty acid composition** or **reduced environmental impact** through altered nitrogen and phosphorus digestion in domestic pigs; avian influenza **virus resistance** in domestic chickens; reduced risk of mastitis in goats and cows; cows producing milk similar to breast milk; and resistance to PRRS virus *(porcine reproductive and respiratory syndrome)* in pigs [42].

However, the largest market for transgenic animals worldwide, including in Germany, concerns research. The share of genetically modified animals in **animal experiments** was around 40% in Germany in 2016. The house mouse may serve as a good example. Thousands of transgenic variants of this mouse can be ordered on the internet. They are used in particular for research into disease models and therapies. From an ethical perspective, the **3R principles** (replacement, reduction and refinement) must be applied. The goal of saving human life with new therapies or medications is a higher-ranking goal than animal protection and is anchored in the fundamental rights of humans. Nevertheless, in accordance with the 3R principles, it must be considered with every experiment whether alternative methods are available [43].

The mouse was the first ever transgenic animal after the German biologist Rudolf Jaenisch, together with the US embryologist Beatrice Mintz, successfully inserted the Simian virus SV40 into the genome of mouse blastocysts in 1974 [44]. In 1985, the first transgenic sheep and pigs were bred for disease research. This was followed in 1997 by a transgenic cow, which was christened Rosie. She was used to test whether it was possible to give cow's milk the properties of human milk. For Europe alone, there are currently more than 1500 patents for genetically modified animals, from mosquitoes to chimpanzees. Since cloning, that is the breeding of genetically identical copies of animals, is no longer a major difficulty (see later in the text), transgenic animals, once produced, can be reproduced comparatively easily.

However, the CRISPR/Cas system will very likely have a lasting impact on breeding, at least indirectly. Since the 1940s, artificial insemination has been of great importance in livestock breeding [45]. Insemination stations, such as the one founded in 1948 in Greifenberg am Ammersee in Bavaria, west of Munich, usually supply several thousand livestock farmers with, for example, cattle semen. The bull's semen is collected and divided into several hundred portions. By selecting high-performance bulls and cows, milk production has quadrupled in the last 40 years. Large-scale artificial insemination is also used in pig and poultry farming [46]. However, it is not equally successful in all breeds. Scientists from the USA have developed a method to stop sperm production in male breeding pigs by eliminating the *NANOS2* gene by gene editing [47]. As a result, these pigs are sterile—but are implanted with sperm-producing stem cells from a breeding animal, a procedure known as **germ cell transplantation** (Fig. 3.11).

Thereupon, these surrogate sires produce semen that is genetically identical to the semen of the stem cell donor. These **"surrogate" sires**, unlike the actual breeding animal, can be cloned well. This means that many twin copies are generated from them. To do this, the fertilized egg is first treated with the CRISPR/Cas system and then several cell divisions are waited for. Then the cell cluster is divided into several. This corresponds to the natural processes in the formation of identical twins. All cell clusters can be transplanted into mothers and carried to term. They are genetically identical. The cloning of the breeding boar could only be done from induced pluripotent stem cells from somatic cells. This is similar to the method used to clone Dolly the sheep in 1996 (Fig. 3.12). However, this method is not yet fully developed and cannot be effectively applied to all animals.

The fact that, despite over 30 years of research into transgenic animal breeding, hardly any products are on the market, or only against great resistance, shows that the 3R principles are also finding their application in animal foods—and by consumers. The fact that the number of vegetarians is currently on the increase suggests that this will not change any time soon. Images of animals barely able to move due to their cultivated obesity are having an effect. However, the aversion is unfortunately still predominantly directed at the use of genetic engineering, quasi as a lightning rod, and not to the same extent at conventional factory farming. This is also reflected in other, ethically less problematic areas of application of genetic engineering. Farmers do see advantages, for example in the resistance of their herds to infections and the resulting possible reduction in antibiotic treatments and vaccinations [48]. Increased meat production, analogous to salmon, is also one of them.

There are old traits, but also numerous traits that are now being implemented using the CRISPR/Cas system (Sect. 5.1). Here, however, at least in the USA, the control system lags behind the state of the art, as gene editing, with the exception of its use in humans, is currently not regulated. It is also important to consider whether gene editing can improve animal welfare by making animals less susceptible to infection and therefore requiring less treatment. And what does it mean for cattle that have literally had their horns taken off with gene editing methods? [49].

Some related societal questions are whether factory farming—and also intensive agriculture—are necessary or substitutable; how will this affect food supply, first to the existing and then to the expected future world population; or how can this be reconciled with the currently existing and expected standards of living (Sect. 3.7).

3.6 Organic Farming and Genetic Engineering

For most people, organic and conventional farming are mutually exclusive. Farming according to organic standards and using genetically engineered seeds or related products then seems all the more incompatible. In their book *Tomorrow's Table*, Pamela Ronald, a US professor of plant pathology, and her husband, Raoul Adamchak, an organic farmer and head of an organic student farm in Davis, California, show that both types of farming—or philosophies?—can indeed be combined [50]. An illustrative example comes from rice farming. Young rice seedlings can survive a few days completely under water. This is used in **organic rice cultivation**, as watering stifles undesirable companion plants. However, it is important to find the right time to drain off the water again, otherwise the rice seedlings will also suffocate. Rice varieties that can survive long periods of flooding are advantageous. Pamela Ronald isolated a gene (called the *SUB1* gene) from an extremely flood-tolerant rice variety called FR13A that she was able to link to this tolerance. When introduced into a low-tolerance rice variety, it was able to survive submerged for 18 days. In addition to rice, the *Sub1* gene has also been successfully tested in wheat, maize and soybean [51–54].

Now the question arises, why not cultivate the tolerant rice variety FR13A instead of isolating the responsible gene and introducing it genetically into other rice varieties. There are two reasons for this. One is yield: the FR13A variety has a low yield. For over 40 years, attempts have been made to cross flood resistance into palatable and high-yielding rice varieties (Fig. 3.13). However, whenever the daughter generation shows flooding-resistant traits, they are accompanied by other, undesirable traits. Thus, the traits appear to be **coupled**.

The second keyword is: varieties. All plants have varieties that are optimally adapted to certain regional **microclimates**. Thus, in plant breeding, desired traits must always be crossed into locally adapted varieties (Fig. 3.13). Therefore, the often-used argument that genetic engineering *per se* reduces biodiversity is not correct. Local varieties and types usually form the basis into which desired genes are crossed. What reduces biodiversity is intensive

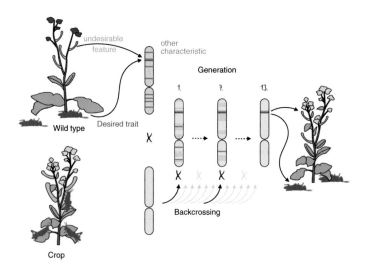

Fig. 3.13 The trait *insecticide tolerance* should be crossed from the red-flowered (wild type) into the yellow-flowered crop. However, as an undesirable trait, the wild type plant has *smaller fruits*. This trait is so close to the desired trait on the chromosome that they cannot be separated—they are coupled. The goal of backcrossing is to cross the desired traits of the wild type into the genetic background of the crop. In the case of crosses from genetically modified plants, this corresponds to the wild type

agriculture *per se*. This does not always succeed, but genetic engineering in general and gene editing in particular (Sect. 5.1) offer a comparatively quick solution here. The two Americans are therefore not the only ones who see an opportunity in the interaction of genetic engineering and organic farming. The renowned *Research Institute of Organic Agriculture* (FiBL), which was founded in Switzerland in 1973, is active in research, development and consultancy for organic agriculture and currently has around 175 employees throughout Europe, also sees opportunities, for example, in the application of gene editing (Sect. 5.1) in organic agriculture. Its former director, Professor Urs Niggli, is one of the world's leading experts in the field of organic farming. He has been advocating a **differentiated discussion** with genetic engineering for quite some time, for which he had to take a lot of criticism from organic associations. In an interview with the daily newspaper *Der Tagesspiegel,* Niggli said:

> After 30 years of argument, it would be desirable to become more objective. [... There are] important problems to solve, such as the fact that we produce more than enough food, but only with vast quantities of pesticides and fertilisers [55].

It is important to question entrenched ideologies and patterns of argumentation, to bring the latest findings into the discussion, to think in colours rather than black and white, and to approach each other when considering opportunities and risks. In a lecture at the University of Vienna in February 2019, Niggli made it clear that if the goal is sustainable agriculture, it is not at all about the question of applying genetic engineering [56]. Rather, one has to take care of appropriate forms of cultivation and land use and move from agribusiness to **agriculture** again. This involves not only an elusive cultural self-image but also very practical action, such as avoiding food waste and using less agricultural land as fodder for livestock farming [57]. Together with other breeding methods (Sect. 3.4), gene editing offers the possibility of introducing traits that are completely different from those considered lucrative by the pharmaceutical giants into plant species that are completely different from those that dominate the world market. Therein lies one of the great opportunities of the CRISPR/Cas system: to become a breeding and development tool for small breeders.

3.7   Risks of Genetic Engineering

Pope Francis writes in the third chapter (The Human Root of the Ecological Crisis), verse 133 of his papal circular *Laudato si'—On Care for Our Common Home* published in June 2015:

> It is difficult to make a general judgement about genetic modification (GM), whether vegetable or animal, medical or agricultural, since these vary greatly among themselves and call for specific considerations. The risks involved are not always due to the techniques used, but rather to their improper or excessive application [58].

In my view, it is immensely important to distinguish between, on the one hand, the risks of technology as such for living beings and the environment and, on the other hand, the social risks of its application or availability. The Pope also makes this distinction at various points in his circular letter. In this section we will deal primarily with the technical risks. And as the epilogue to my preface already indicated: science has been aware of them from the very beginning. Let us first take a look back at the beginnings of genetic engineering.

In 1973, aspects of the first genetic engineering work by Boyer and Cohen (Sect. 3.4) were presented at the renowned *Gordon Research Conference* in the USA and their implications were immediately recognised: namely, the possibility of recombining DNA from different organisms and thus creating hybrid organisms. The research recognised both opportunities and risks. I would almost call it the *Sermon on the Mount*, the letter published in 1974 by the US biochemist Paul Berg and ten other scientists in the scientific journal *Science,* which warns of the possible risks in dealing with recombinant DNA [59]. In the letter they demanded firstly a halt of all work with recombinant DNA, secondly a careful review of planned experiments, thirdly the installation of a national advisory committee and fourthly the convening of an international scientific conference to discuss the future handling of the new possibilities. The authors write that *our concern is based on the assessment of potential and undetectable risks.* These were thus precautionary measures proposed by the authors themselves.

This is a living example of the application of the precautionary principle, which is also the basis of the European genetic engineering law. In contrast to the **scientific principle,** according to which only proven hazards are considered, the **precautionary principle** is based on a risk or hazard assessment in advance. As a result, there was an eight-month moratorium from July 1974, that is all work on recombinant DNA was suspended. In February 1975, the important four-day *Asilomar Conference on Recombinant DNA Molecules took* place in Asilomar, California [60, 61]. For the first time, a scientific conference was attended not only by researchers, but also by government personnel and representatives of the judiciary and journalism. Based on the discussions during the conference, the US National Institute of Health (NIH) was commissioned to draw up general valid safety guidelines. These were published in

1976 as *Guidelines for the Evaluation of Work with Recombinant DNA and Recombinant Organisms* [62]. These set out four levels of safety for the physical and biological safeguarding of experiments and organisms. These **safety levels (S1-S4)** are still valid today and serve to protect humans and the environment. According to the German **Genetic Engineering Act**:

Safety level 1 for genetic engineering work that is not expected to pose a risk to humans or the environment. This also includes the experimental transfer of DNA fragments between bacteria of the same species (**self-cloning**). This work does not require a permit, but it does require notification and documentation. In addition, appropriate expertise must be available, which is usually demonstrated by an academic degree and two years of work experience. Many schools now have S1 laboratories. In the U.S., S1 work is deregulated, so it can be done by anyone, anywhere.

Safety level 2 for genetic engineering work where a low risk to humans and the environment can be assumed. This includes work with many pathogens, such as the measles viruses.

Safety level 3 for genetic engineering work where a moderate risk to humans and the environment can be assumed. This includes, for example, the HI virus, the trigger of the immune deficiency **AIDS**.

Safety level 4 for genetic engineering work where there is a high risk or reasonable suspicion of such a risk to humans and the environment, for example when working with **Ebola viruses**. There are four S4 laboratories in Germany: at the *Bernhard Nocht Institute for Tropical Medicine* in Hamburg, at the *Institute for New and Novel Animal Pathogens* (INNT) of the *Friedrich Loeffler*

Institute (FLI) on the island of Reims, at the *Institute of Virology* of the Philipps University in Marburg and at the *Robert Koch Institute* (RKI) in Berlin.

For employees, laboratories, greenhouses, animal husbandry rooms or production facilities, protective devices and measures corresponding to the safety levels apply—incidentally, these also apply when non-genetically engineered work is being carried out. While these safety levels are still valid today, the **Cartagena Protocol** on Biosafety in particular regulates cross-border trade in *living genetically modified organisms*. Products of genetically modified organisms are explicitly excluded. The protocol came into force in September 2003 and 171 countries have signed it so far, but not, for example, the USA, Canada, Argentina or Australia.

But how great is the risk that is actually being regulated here? First of all, it is important to distinguish risks from hazards. **Risk** is understood as the possibility of damage occurring. It comprises the product of the extent and probability of occurrence of a harm. For example, there is a risk that when I boil an egg, I unintentionally get a hard boiled. However, there is no hazard associated with this risk. A **hazard** exists when a harmful effect is caused to a target organism or the environment. This requires a mechanism of action. Many risks and hazards attributed to genetic engineering, especially in agriculture, exist even if genetic engineering is not used. For example, the reduction of biodiversity (e.g. through monocultures or the use of pesticides), the exploitation of farmers, especially in developing countries (e.g. by tying them to cultivation licences), the spread of antibiotic resistance (through faecal waste from animal husbandry, hospitals, etc., or old transgenic plant varieties) or the displacement of native species (worldwide import and export as well as release of living organisms; spread via water ballast in shipping) have many causes. Nevertheless, the assessment of risks and hazards is extremely important, especially in the fields of nutrition and medicine. However, this also requires established risk research to gather criteria for assessment. The fact that no more field trials with GMOs have taken place in Germany since 2013 also shows that no more safety research is taking place in this field. But a greenhouse cannot simulate an open field. At the *Leibniz Institute of Plant Genetics and Crop Plant Research* (IPK) in Gatersleben, a plant culture hall was opened in August 2017 in which environmental conditions can be simulated at great expense. The costs amounted to over eight million euros. Added to this are immense operating and energy costs. In such a hall, GMOs can also be exposed to a wide variety of simulated environmental conditions as part of risk research. What still happens far too little is the comprehensive weighing of risks. Here, for example, between the release of GMOs and the consequences of the energy-intensive and thus climate-damaging simulation.

It is also important to distinguish whether a risk relates to the technique used or to its outcome, which could equally have been achieved by other means. To put it more concretely, results of transgenetic engineering can hardly be achieved with modern breeding if the source and target organism of the transgenes do not meet naturally. But conventional breeding can also lead to problems in the environment as a result. A good example is provided by the cultivation of rapeseed. Rapeseed oil was already in demand in the Middle Ages as a lubricant and lamp oil. It was not to be found in salads because the harmful erucic acid and bitter glucosinulates from the rape seeds made the oil inedible—incidentally also for deer, hares and other wild animals. The field was safe. It is thanks to the success of selection breeding (including that of my namesake Professor Gerhard Röbbelen of Göttingen University) that erucic acid-free varieties, so-called 0-varieties, have been cultivated since 1974, and since 1986 additionally glucosinolate-poor 00-varieties [63, 64]. This also tastes good to the wild animals, however, the consumption can lead to digestive disorders and death in them due to the high protein content.

What are the possible **risks** of genetic engineering? On the one hand, **unexpected genetic changes** can occur in genetically modified organisms. In the first experimental release of genetically modified petunias approved for research purposes in Germany at the *Max Planck Institute for Plant Breeding Research* in Cologne in May 1990, unexpected behaviour occurred. A gene from maize was transferred to the 30,000 petunias released, which changed the flower colour from white to salmon red [65]. In addition, a so-called *jumping gene* was inserted into the plants, causing random mutagenesis. This undirected method is known as **transposon mutagenesis**. If the jumping gene were to insert itself into the flower color gene, its function would be disrupted. This should only happen extremely rarely and would be recognizable by the fact that the flower colour would then be white or red-white speckled again. To the great surprise of the scientists, however, over 60% of the flowers were white or speckled. Opponents of genetic engineering saw this as a failure of science because the experiment had got out of control and genetic engineering was obviously not controllable. This is to be countered by the fact that this is precisely what experiments have to be carried out for. At no time did the plantings pose a risk to humans or the environment. Further investigations showed that **environmental factors** such as UV radiation have an effect on the expression of the genes—an important finding. A similar unexpected behaviour was also observed in petunias in another experiment in 1990. In this experiment, US scientists inserted several copies of a gene that determines flower colour (chalcone synthase) into the petunia genome. The hope was that multiple copies would lead to increased activity, in this case to a stronger flower colour. This was also observed—but it was observed in many more plants that the enzyme was no longer active at all, so the flowers were white [66]. In the following years, similar observations were also made in fungi and animals. A detailed investigation of the underlying molecular biological mechanism led to the award of the Nobel Prize in 2006 to the American scientists Andrew Fire and Craig Mello: They were able to show that double-stranded RNA is degraded in cells by a specific process. This process is now known as **RNAi** (RNA interference) and is now widely used technically as a gene regulatory method [67]. In the case of chalcone synthase in petunia, whether the RNA transcripts can form a double strand (i.e., interfere) depends on the orientation of the genes in the genome. If they both point in the same direction, the desired increased expression occurs because twice as many transcripts are present. If, on the other hand, they are oriented in opposite directions, interference occurs, that is the formation of double-stranded RNA. As a result, it is degraded and no chalcone synthase can be formed—the flowers are white. Both events from the early days of GMO release have in common that the genetic constructs did not behave as expected. Investigation of the causes helped to improve the methodology and was not associated with any danger.

A major concern for many is the **unintentional release** of genetically modified organisms. Finnish molecular biologist Teemu Teeri first noticed orange petunias blooming in flower pots at Helsinki railway station in the summer of

2015 [68]. He knew that petunias do not naturally bloom orange and wanted to know what was going on. The city's gardeners told him it was the **Bonnie Orange** variety and gave him some plants. When he examined samples of the petunias in his lab, he found the *A1* gene from corn. It codes for the enzyme dihydroflavonol-4-reductase, which influences flower colour and was also used in investigations at the *Max Planck Institute for Plant Breeding Research* in Cologne. In addition, he and his team found, among other things, the *nptII* marker gene, which codes for antibiotic resistance and was previously widely used in the production and selection of GMOs. There was no doubt in Teeri's mind that he was looking at an unlabelled transgenic petunia. Further genetic analysis of petunias showed that numerous unlabelled **transgenic petunia varieties** were on the market in Finland and Europe. In April 2017, the Finnish Food Safety Authority informed that eight transgenic petunia varieties could be found and commercially used in Finland without approval. The plants came from companies in the Netherlands and Germany. Subsequently, 51 different orange and salmon petunia varieties have been described worldwide that are transgenic, unregistered and in commercial use. How could this happen? The genetic construct developed for research purposes at the *Max Planck Institute for Plant Breeding Research* in Cologne (a construct was also found which was proven to originate from China) was apparently used by a Dutch breeding company for breeding new varieties after the acquisition of the licensing rights. In 1995 it was presented as ready for the market [69]. However, an application for commercial cultivation was never submitted, as presumably the financial outlay for approval under genetic engineering law was too high. Consciously or unconsciously, the transgenic construct got into commercial petunia varieties and was probably in commerce worldwide by the late 1990s. In 2017, a worldwide **recall campaign** was launched, which soon also affected farms in Germany—here, the orange flowering petunia could be found in several garden departments of wholesale chains.

Unintentional releases of genetically modified crops are reported time and again. As recently as November 2018, France informed the EU Commission that very small amounts of a genetically modified rapeseed variety (GT73) not approved for cultivation had been detected in seed. This variety, developed by *Monsanto* (now *Bayer*), is resistant to glyphosate. New seed was generated from this contaminated seed and sold throughout Europe. As a result, seeds were sown on agricultural land in 84 farms and on about 21 square kilometres in ten German states. The contamination amounted to about 0.1%. The good news is that all of this has been accurately tracked and farms have been asked to plough up the plants that have not yet flowered. In addition, the manufacturer has compensated the farms for the crop damage incurred.

The horizontal transfer and spread of genes from species to species is also a frequently mentioned risk. **Horizontal gene transfer** is well known from bacteria, for example. It is an essential mechanism that leads to genetic variability and thus adaptation to very different environments. For example, a single intestinal bacterium of the species *Escherichia coli* encodes around 2000–4000 genes. In this alone, the strains differ. However, the total of all gene variants in all known *Escherichia coli* genomes worldwide, the so-called **pangenome**, amounts to about 18,000 genes [70, 71]. Genomes are therefore much more diverse (more plastic) and also more changeable over time (more dynamic) than is generally assumed. And so, bacteria are a good vehicle for genes. Some time ago, **antibiotic resistance genes** against man-made, artificial antibiotics were discovered, for example, in bacteria of birds in the Antarctic or of indigenous people in the Amazon basin [72–74]. It can therefore be assumed in principle that parts of the genetic material of a released GMO, after it dies and rots, can be taken up by bacteria, integrated into the bacterial genome and spread globally. The likelihood of dissemination becomes especially high if the genetic information gives an advantage to the founder population. Otherwise, the genetic ballast is more likely to be discarded. This is also true for plants and animals—without a selection advantage, a new genetic information (allele) is unlikely to spread and be maintained in a population. Via gene shuttles such as the agrobacterium in plants or viruses in animals and plants, DNA fragments can in principle also be transferred and spread from the bacterial world to the animal and plant world. The probabilities are very low and the dangers must be assessed on a **case-by-case basis.** While the risk of spreading antibiotic resistance genes is obvious, it is more difficult to imagine in the case of spreading a rice gene for flooding resistance (Sect. 3.6). However, since January 2019, horizontal gene transfer has been given expanded significance. Analysis of the genome of cockatoo grass *(Alloteropsis semialata),* which grows in tropical and subtropical countries, revealed 57 genes originating from at least nine different donor grasses of different species [75]. Thus, there has been a transfer of these genes across species boundaries. Some of these transfers took place about two million years ago—so on average there was one gene transfer in about 35,000 years. It is thought that the horizontally incorporated genes have a role in resistance to environmental factors and disease, as well as in increasing the efficiency of the conversion of sunlight into chemical energy (photosynthesis). Exactly what the molecular mechanisms of gene uptake are has not yet been described.

What is clear is that by taking up the foreign genes, the cockatoo grass expands its molecular toolbox and is thus better adapted to the environment, presumably giving itself an evolutionary advantage. However, the transfer of about 60 genes in two million years does not exactly indicate a very active process. Incidentally, the exchange of genetic material on the fused tissue has also been observed in grafting (the grafting of ornamental cacti and woody plants such as fruit trees or grapevines) [76].

Time and again, and not just since the use of genetic engineering, there have been warnings about the possible development of **resistance**. Insects can become resistant to active substances in Bt plants (Sect. 3.4), weeds can develop resistance to plant destruction agents such as glyphosate. Clueless are those who hope to prevent this. In evolutionary biology, there is a hypothesis called the ***Red Queen Hypothesis***. Formulated in 1973 by the U.S. evolutionary biologist Leigh van Valen, it states that sexual reproduction and the rate of the resulting rearrangement of genetic information in a species are just enough to adapt to an ever-changing environment [77]. That environment, of course, includes other and competing creatures. And thus, the *Red Queen Hypothesis* ascribes to an engine of evolution that consists in constant change and resulting adaptation. Living things must constantly develop new, or adapted, mechanisms to protect themselves against parasites, for example, and to avoid extinction. Van Valen uses for his hypothesis the image of the Red Queen from Lewis Carroll's story *Alice behind the Mirrors* from 1871 [78]. In the *Garden of Talking Flowers*, Alice meets the Queen in a landscape like a chessboard. Accordingly, the Red Queen is translated in German as the Black Queen. She takes Alice by the hand and runs with her through the forest at the greatest possible speed, without moving from the spot. When Alice notices this, the Black Queen responds:

In this country you have to run as fast as you can if you want to stay in the same place. And to get somewhere else, you have to run at least twice as fast!

The *Red Queen Hypothesis* describes an observation that is probably familiar to every reader and that Reinhold Messner summarized with *speed is safety*: Dynamic behaviour increases responsiveness. One is familiar with this in boxing in the ring or fencing on the planche. Athletes are always bobbing and prancing to prepare or ward off the next attack. This is also known from metabolism. Higher turnover rates in enzymes allow faster adaptation to changes in the cellular environment [79]. So, it is an illusion to assume that there is a way to avoid resistance formation. It is part of being alive. I call this the **garlic effect**: it is what stinks that works. To counteract and delay the development of resistance, balanced drug management is essential. For example, the use of the same active ingredient over a period of years should be avoided. In addition, good arable farming practices should again be given increasing importance. Mixed crops and varied crop rotations can make an important contribution here.

The **indirect risks** seem to me to be much greater than the direct risks to ecosystems and the environment. For example, herbicide-resistant plants tempt us to apply too much rather than too little plant protection product, such as **glyphosate**, to the fields. Higher concentrations may also be used. This can have not only health effects on farmers, but also long-term effects on the environment. For example, glyphosate is considered a so-called chelating agent. This means that glyphosate can bind ions in groundwater and thus impair groundwater quality. Holistic thinking is required here.

A few years ago, the so-called **Séralini study**, named after the French molecular biologist Gilles-Eric Séralini, became highly publicized. In a study published in September 2012, his research team described how rats fed the genetically modified maize variety NK603 developed cancer and died earlier [80]. NK603 is a glyphosate-resistant corn variety made by *Monsanto* (now *Bayer*). This came as a bombshell and was picked up and spread by the media—all the way to the daily news. Both the EU Commission and the German government had their scientific committees investigate the study results. Criticism quickly followed, concerning both the design of the study with too few test animals and the selection of the test animals. The breed used was prone to tumour formation anyway. There were also opposing voices, which in turn accused the critics of using unequal measures: *Monsanto*'s own studies, which led to the import approval of the corn variety in Europe, were also based on the same breed of rats. In November 2013, the journal *Food and Chemical Toxicology* retracted the Séralini study on NK603 corn as scientifically untenable. From April 2014 to 2018, an international team of researchers with scientists from six countries repeated the Séralini study experiment on behalf of the EU Commission. The results were published in February 2019 and concluded that *no adverse effects* could be observed [81]. In an assessment of the available scientific literature on feeding studies, a 2017 review concluded that adverse effects were reported in 5% of the studies [82]. However, all of these studies had methodological weaknesses and had appeared in comparatively unknown scientific journals.

The use of genetic engineering in **gene therapy** may pose an immediate hazard (see the Jesse Gelsinger case in Sect. 5.3). This is due not least to the fact that, for ethical reasons, no large-scale clinical studies are available and, in the case of application in the germ line, such studies are prohibited anyway. Epigenetic effects are also still largely not understood (Sect. 8.1). In the case of somatic gene therapy, but especially in the case of germline gene therapy, very careful consideration must be given: How great is the risk of unexpected effects occurring, accompanied by the danger of not being able to treat an acute or expected disease (Sect. 5.3). We must consider a further potential hazard in the application of genetic engineering by **laypersons** (Sect. 7.1). The use of any technique without underlying experience usually poses a hazard. For example, programmed computer viruses have caused damage all over the world, sometimes out of playful impulse. Depending on one's personality, a minor hobby may turn into an endeavour to be publicly noticed. In such hands, modern genetic engineering methods (Sect. 6.2), computer viruses as it were, can be used to design harmful viruses or living beings—whether intentionally or unintentionally. How this hazard can be controlled beyond official regulations and *community ethics* (self-imposed standards of biohackers) remains open (Sect. 7.1) [83].

Gene editing is currently bringing a great deal of dynamism to the risk debate (see in particular Sect. 5.1 and Sect. 5.2). At this point, I would like to refer to a special form of application of gene editing (Sect. 5.1), the **gene drive**. Here, the genes for the CRISPR/Cas system are inserted into the genome of an organism, that is a genetically modified organism is produced. The organism is then able to act on its own genome and, after fertilization, on the "new" genome. Thus, the organism itself becomes a genetic engineer, so to speak. This ability is passed on to subsequent generations. For example, work is being done on mosquitoes that are immune to **malaria** and thus cannot infect humans. By means of the gene-drive system, they pass on this immunity to wild, non-immune mosquitoes with which they interbreed (Fig. 3.14). Thus, the system causes

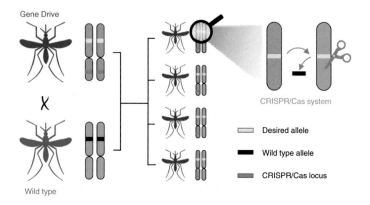

Fig. 3.14 Gene drive bypasses the Mendelian mode of inheritance. Instead of a mix of maternal and paternal alleles, the gene drive leads to the wild-type allele being displaced from the population. This happens because the gene drive mosquito also carries the genes for the CRISPR/Cas system in its genome

the gene-drive mosquitoes to genetically modify other mosquitoes. The immunity spreads through the population. This is referred to as a mutagenic chain reaction or **super-Mendelian inheritance**. In Mendelian inheritance, the offspring receive one set of chromosomes from the father and one from the mother. Thus, they have one paternal and one maternal allele of each gene (Fig. 4.6). In gene drive, things are different. The gene-drive allele overwrites the other copy. Therefore, all offspring carry the same allele twice, that is, only the desired variant of a gene. The same is true, of course, for all subsequent generations. In January 2019, it was shown that gene drive also works in mice [84]. It must be clearly stated that a system is being created here that cannot be stopped [85]. Currently, research is being done to see if some sort of molecular switch can be added to the drive [86]. This might make it possible to regulate it. In the case of transgenic organisms, a deliberate attempt is usually made to prevent the spread of transgenes in nature. The gene drive has exactly the opposite goal [87].

Based on the gene drive, research is currently being carried out at the US Pentagon's research institute into a way of genetically modifying crops in the field [88, 89]. Insects are to infect the plants with genetically modified viruses. These viral vectors would in turn contain the genes for the CRISPR/Cas system that modify the plant genomes. Based on the current state of knowledge, the application of such technologies must be viewed very critically. Regardless of whether this system would additionally be equipped with a gene drive or not, retrievability is just as difficult to imagine here as controlling the spread. This method also has a high potential for misuse. Therefore, research in this area is absolutely necessary, but an active application in the field is associated with high risks and dangers according to the current state of knowledge.

Ultimately, the societal risks associated with unequal **participation** in genetic engineering must also be evaluated. Who will be able to afford a gene therapy, a high-yield crop, an engineered biodiesel-from-sunlight bacterium, or a novel drug developed with the help of genetic engineering? These questions extend beyond the scope of the life sciences. I cannot present any solutions here. However, I share the conviction of many scientists that we should be open to genetic engineering—as a technology. Even after more than a quarter of a century of its application, we cannot identify any greater risks than those already posed by the areas in which genetic engineering is used.

However, risks must always be reassessed. That is why risk research is an important sector that a highly technological society must afford. After all, the risks of a technology can only ever be assessed according to the current state of knowledge. In addition, it is necessary to constantly reassess existing data and include new data. For example, the formation of the ozone hole over Antarctica was overlooked for a long time because the data from NASA's TOMS (*Total Ozone Mapping Spectrometer*) satellites were incorrectly evaluated over many years—by a computer program. This program marked ozone values that were too low as errone-ous and these were thus excluded from the data analysis [90]. A new examination of the data in the 1980s, based on a publication by British scientists, showed that the program had flagged too much data as erroneous and thus filtered it out [91]. The ozone hole could have been detected years earlier.

References

1. Bacon F (2009) The New Atlantis. Dodo Press, London/UK
2. Darwin C (2019) On the origin of species. Puffin, London/UK
3. Mendel G (1866) Versuche über Pflanzen-Hybriden. Verhandlungen des Naturforschenden Vereins zu Brünn 4: 3–47
4. de Vries, H (1901–03) Die Mutationstheorie. Bde 1 u 2. Verlage von Veit & Co., Leipzig. doi:https://doi.org/10.5962/bhl.title.11336
5. Malling HV (2004) History of the science of mutagenesis from a personal perspective. Environ Mol Mutagen 44: 372–386. doi:https://doi.org/10.1002/em.20064
6. Mishra R (ed) (2012) Mutagenesis. InTech, Rijeka/HR. doi:https://doi.org/10.5772/2937
7. Muller HJ (1927) Artificial Transmutation of the gene. Science 66: 84–87. doi:https://doi.org/10.1126/science.66.1699.84
8. Stadler LJ (1928) Mutations in barley induced by x-rays and radium. Science 68: 186–187. doi:https://doi.org/10.1126/science.68.1756.186
9. Vose PB (1980) Introduction to nuclear techniques in agronomy and plant biology. Pergamon Press Ltd, Oxford/UK

10. Johnson P (2013) Safeguarding the atom: the nuclear enthusiasm of Muriel Howorth. Brit J Hist Sci 45: 551–571. doi:https://doi.org/10.1017/S0007087412001057

11. Murray MJ, Todd WA (1972) Registration of Todd's Mitcham Peppermint. Crop Sci 12: 128. doi:https://doi.org/10.2135/cropsci1972.0011183x001200010056x

12. Broertjes C, van Harten AM (Edt) (1988) Applied mutation breeding for vegetatively propagated crops: Vol 12. Elsevier Science Ltd, Amsterdam/NL

13. Case 528/16. (2018) In: InfoCuria. Visited 23.04.2019: curia.europa.eu/juris/documents.jsf?num=c-528/16

14. Nabholz M, Miggiano V, Bodmer W (1969) Genetic analysis with human-mouse somatic cell hybrids. Nature 223: 358–363. doi:https://doi.org/10.1038/223358a0

15. ISAAA (2017) Global Status of Commercialized Biotech/GM Crops in 2017: Biotech Crop Adoption Surges as Economic Benefits Accumulate in 22 Years. ISAAA Brief No. 53. ISAAA, Ithaca, New York/USA

16. Anton AC, Grommen R, Hessels G, et al (2012) Studie über einen neuen potenziellen Impfstoff gegen *Rhodococcus equi*-Infektionen bei Fohlen. Tierärztliche Umschau 67:394–400

17. Lederberg J (1952) Cell genetics and hereditary symbiosis. Physiol Rev 32: 403–430. doi:https://doi.org/10.1152/physrev.1952.32.4.403

18. Chang AC, Cohen SN (1974) Genome construction between bacterial species in vitro: replication and expression of *Staphylococcus* plasmid genes in *Escherichia coli.* Proc Natl Acad Sci USA 71: 1030–1034. doi:https://doi.org/10.1073/pnas.71.4.1030

19. Cohen SN, Boyer HW (1980) Process for Producing Biologically Functional Molecular Chimeras. US Patent 4,237,224, initially filed November 4, 1974, issued December 2, 1980

20. Berg P, Mertz JE (2010) Personal Reflections on the Origins and Emergence of Recombinant DNA Technology. Genetics 184: 9–17. doi:https://doi.org/10.1534/genetics.109.112144

21. Schell J, Van Montagu M (1977) The Ti-Plasmid of *Agrobacterium tumefaciens,* A Natural Vector for the Introduction

of NIF Genes in Plants? In: Genetic Engineering for Nitrogen Fixation. Springer, Boston, Massachusetts/USA, S. 159–179. doi:https://doi.org/10.1007/978-1-4684-0880-5_12

22. Zambryski P, Joos H, Genetello C, et al (1983) Ti plasmid vector for the introduction of DNA into plant cells without alteration of their normal regeneration capacity. EMBO J 2: 2143–2150. doi:https://doi.org/10.1002/j.1460-2075.1983. tb01715.x

23. Muth J, Hartje S, Twyman RM, et al (2008) Precision breeding for novel starch variants in potato. Plant Biotechnol J 6: 576–584.   doi:https://doi.org/10.1111/j.1467-7652.2008. 00340.x

24. Andersson M, Turesson H, Olsson N, et al (2018) Genome editing in potato via CRISPR-Cas9 ribonucleoprotein delivery.   Physiol   Plant   164:   378–384. doi:https://doi. org/10.1111/ppl.12731

25. Callaway E (2018) CRISPR plants now subject to tough GM laws in European Union. Nature 560: 16. doi:https://doi. org/10.1038/d41586-018-05814-6

26. Eckerstorfer MF, Engelhard M, Heissenberger A, et al (2019) Plants Developed by New Genetic Modification Techniques— Comparison of Existing Regulatory Frameworks in the EU and Non-EU Countries. Front Bioeng Biotechnol 7: 26. doi:https://doi.org/10.3389/fbioe.2019.00026

27. Kahrmann J, Leggewie G (2018) Gentechnikrechtliches Grundsatzurteil des EuGH und die Folgefragen für das deutsche Recht. NuR 40: 761–765. doi:https://doi. org/10.1007/s10357-018-3429-8

28. Bulla LA (1975) Bacteria as insect pathogens. Annu Rev Microbiol 29: 163–190. doi:https://doi.org/10.1146/annurev.mi.29.100175.001115

29. Candas M, Bulla LA (2002) Microbial insecticides. In: Bitton G (Ed) Encyclopedia of Environmental Microbiology. John Wiley and Sons, New York/USA, S 1709-17. doi:https://doi. org/10.1002/0471263397.env258

30. Franz JM, Krieg A (1961) Schädlingsbekämpfung mit Bakterien (*Bacillus thuringiensis*). Gesunde Pflanzen 13: 99–204

31. Bioland Richtlinien (2019) Bioland e. V., Mainz
32. Gonsalves D, Ferreira S (2003) Transgenic Papaya: A Case for Managing Risks of Papaya ringspot virus in Hawaii. Plant Health Progress 4: 17. doi:https://doi.org/10.1094/php-2003-1113-03-rv
33. Tena G (2017) Sweet transgenic immunity. Nat Plants 3: 911. doi:https://doi.org/10.1038/s41477-017-0080-y
34. Dale J, James A, Paul J-Y, et al (2017) Transgenic Cavendish bananas with resistance to Fusarium wilt tropical race 4. Nature Comm 8: 1496. doi:https://doi.org/10.1038/s41467-017-01670-6
35. Pew Research Center (2018) Most Americans Accept Genetic Engineering of Animals That Benefits Human Health, but Many Oppose Other Uses. In: Pew Research Center. Aufgerufen am 16.03.2019: pewinternet.org/wp-content/uploads/sites/9/2018/08/PS_2018.08.16_biotech-animals_FINAL.pdf
36. Editorial (2019) Hybrid embryos, ketamine drug and dark photons. Nature 567: 150–151. doi:https://doi.org/10.1038/d41586-019-00790-x
37. Ledford H (2015) Salmon approval heralds rethink of transgenic animals. Nature 527: 417–418. doi:https://doi.org/10.1038/527417a
38. Kay MA (2011) State-of-the-art gene-based therapies: the road ahead. Nat Rev Genet 12: 316–328. doi:https://doi.org/10.1038/nrg2971
39. Gordon JW, Ruddle FH (1981) Integration and stable germ line transmission of genes injected into mouse pronuclei. Science 214: 1244–1246. doi:https://doi.org/10.1126/science.6272397
40. Smith K, Spadafora C (2005) Sperm-mediated gene transfer: Applications and implications. BioEssays 27: 551–562. doi:https://doi.org/10.1002/bies.20211
41. Dobrinski I (2005) Germ Cell Transplantation. Semin Reprod Med23:257–265.doi:https://doi.org/10.1055/s-2005-872454
42. Yang B, Wang J, Tang B, et al (2011) Characterization of Bioactive Recombinant Human Lysozyme Expressed in Milk

of Cloned Transgenic Cattle. PLoS One 6: e17593. doi:https://doi.org/10.1371/journal.pone.0017593

43. Diekämper J, Fangerau H, Fehse B, et al (eds) (2018) Vierter Gentechnologiebericht. Nomos Verlagsgesellschaft, Baden-Baden

44. Jaenisch R, Mintz B (1974) Simian virus 40 DNA sequences in DNA of healthy adult mice derived from preimplantation blastocysts injected with viral DNA. Proc Natl Acad Sci USA 71: 1250–1254. doi:https://doi.org/10.1073/pnas.71.4.1250

45. Manafi M (Ed) (2011) Artificial Insemination in Farm Animals. InTech, Rijeka/HR. doi:https://doi.org/10.5772/713

46. Schramm GP, Nutztieren HPKBB, 1991 (2005) Künstliche Besamung beim Geflügel. Züchtungskunde 77: 206–217

47. Park K-E, Kaucher AV, Powell A, et al (2017) Generation of germline ablated male pigs by CRISPR/Cas9 editing of the *NANOS2* gene. Sci Rep 7: 40176. doi:https://doi.org/10.1038/srep40176

48. Tait-Burkard C, Doeschl-Wilson A, McGrew MJ, et al (2018) Livestock 2.0—genome editing for fitter, healthier, and more productive farmed animals. Genome Biol 19: 204. doi:https://doi.org/10.1186/s13059-018-1583-1

49. Carlson DF, Lancto CA, Bin Zang, et al (2016) Production of hornless dairy cattle from genome-edited cell lines. Nat Biotechnol 34: 479–481. doi:https://doi.org/10.1038/nbt.3560

50. Ronald PC, Adamchak RW (2008) Tomorrow's Table: Organic Farming, Genetics, and the Future of Food. Oxford University Press, New York/USA. doi:https://doi.org/10.1093/acprof:oso/9780195301755.001.0001

51. Xu K, Xu X, Fukao T, et al (2006) Sub1A is an ethylene-response-factor-like gene that confers submergence tolerance to rice. Nature 442: 705–708. doi:https://doi.org/10.1038/nature04920

52. Mackill DJ, Ismail AM, Singh US, et al (2012) Development and Rapid Adoption of Submergence-Tolerant (*Sub1*) Rice Varieties. Adv Agron 115: 299–352. doi:https://doi.org/10.1016/B978-0-12-394276-0.00006-8

53. Herzog M, Fukao T, Winkel A, et al (2018) Physiology, gene expression, and metabolome of two wheat cultivars with contrasting submergence tolerance. Plant, Cell Environ 41: 1632–1644. doi:https://doi.org/10.1111/pce.13211

54. Herzog M (2017) Mechanisms of flood tolerance in wheat and rice. University of Copenhagen, Dissertation

55. Karberg S (2018) Crispr ist nicht immer Gentechnik. In: Der Tagesspiegel. Aufgerufen am 16.04.2019: tagesspiegel.de/wissen/europaeischer-gerichtshof-vor-der-entscheidung-crispr-ist-nicht-immer-gentechnik/20864058.html

56. Universität für Bodenkultur Wien (2019) Biolandbau und Gene Editing—eine (un-)mögliche Kombination? In: YouTube. Aufgerufen am 27.02.2019: youtu.be/mGhV0BvXnsg

57. Muller A, Schader C, Scialabba NE-H, et al (2017) Strategies for feeding the world more sustainably with organic agriculture. Nature Comm 8: 1290. doi:https://doi.org/10.1038/s41467-017-01410-w

58. Papst Franziskus (2015) Enzyklika Laudato Si'. Libreria Editrice Vaticana

59. Berg P, Baltimore D, Boyer HW, et al (1974) Potential biohazards of recombinant DNA molecules. Science 185: 303. doi:https://doi.org/10.1126/science.185.4148.303

60. Berg P, Baltimore D, Brenner S, et al (1975) Summary statement of the Asilomar conference on recombinant DNA molecules. Proc Natl Acad Sci USA 72: 1981–1984. doi:https://doi.org/10.1073/pnas.72.6.1981

61. Berg P (2008) Meetings that changed the world: Asilomar 1975: DNA modification secured. Nature 455: 290–291. doi:https://doi.org/10.1038/455290a

62. Gartland WJ, Stetten D (1976) Guidelines for Research Involving Recombinant DNA Molecules. National Institutes of Health (U.S.)

63. Röbbelen G (1976) Züchtung und Erzeugung von Qualitätsraps in Europa. Eur J Lipid Sci Technol 78: 10–17. doi:https://doi.org/10.1002/lipi.19760780102

64. Sauermann W (2014) Von 0 auf 00 bis zum Hybridraps. Bauernblatt 28–31

65. Meyer P, Heidmann I, Forkmann G, Saedler H (1987) A new petunia flower colour generated by transformation of a mutant with a maize gene. Nature 330: 677–678. doi:https://doi.org/10.1038/330677a0

66. Napoli C, Lemieux C, Jorgensen R (1990) Introduction of a chimeric chalcone synthase gene into petunia results in reversible co-suppression of homologous genes in trans. Plant Cell 2: 279–289. doi:https://doi.org/10.1105/tpc.2.4.279

67. Sen GL, Blau HM (2006) A brief history of RNAi: the silence of the genes. FASEB J 20: 1293–1299. doi:https://doi.org/10.1096/fj.06-6014rev

68. Bashandy H, Teeri TH (2017) Genetically engineered orange petunias on the market. Planta 246: 277–280. doi:https://doi.org/10.1007/s00425-017-2722-8

69. Oud JSN, Schneiders H, Kool AJ, van Grinsven MQJM (1995) Breeding of transgenic orange *Petunia hybrida* varieties. In: The Methodology of Plant Genetic Manipulation: Criteria for Decision Making. Springer Verlag, Dordrecht/NL, S 403–409. doi:https://doi.org/10.1007/978-94-011-0357-2_49

70. Touchon M, Hoede C, Tenaillon O, et al (2009) Organised Genome Dynamics in the *Escherichia coli* Species Results in Highly Diverse Adaptive Paths. PLoS Genet 5: e1000344. doi:https://doi.org/10.1371/journal.pgen.1000344

71. Stokes HW, Gillings MR (2011) Gene flow, mobile genetic elements and the recruitment of antibiotic resistance genes into Gram-negative pathogens. FEMS Microbiol Rev 35: 790–819. doi:https://doi.org/10.1111/j.1574-6976.2011.00273.x

72. Sjölund M, Bonnedahl J, Hernandez J, et al (2008) Dissemination of Multidrug-Resistant Bacteria into the Arctic. Emerging Infect Dis 14: 70–72. doi:https://doi.org/10.3201/eid1401.070704

73. Bartoloni A, Pallecchi L, Rodríguez H, et al (2009) Antibiotic resistance in a very remote Amazonas community. Int J Antimicrob Agents 33: 125–129. doi:https://doi.org/10.1016/j.ijantimicag.2008.07.029

74. Clemente JC, Pehrsson EC, Blaser MJ, et al (2015) The microbiome of uncontacted Amerindians. Sci Adv 1: e1500183. doi:https://doi.org/10.1126/sciadv.1500183

75. Dunning LT, Olofsson JK, Parisod C, et al (2019) Lateral transfers of large DNA fragments spread functional genes among grasses. Proc Natl Acad Sci USA 116: 4416–4425. doi:https://doi.org/10.1073/pnas.1810031116

76. Stegemann S, Bock R (2009) Exchange of Genetic Material Between Cells in Plant Tissue Grafts. 324: 649–651. doi:https://doi.org/10.1126/science.1170397

77. Van Valen L (1973) A new evolutionary law. Evol Theory 1: 1–30

78. Carroll L (1974) Alice hinter den Spiegeln. Insel Verlag, Leipzig

79. Hochachka PW, Somero GN (2002) Biochemical Adaptation. Oxford University Press, New York/USA

80. Séralini G-E, Clair E, Mesnage R, et al (2012) RETRACTED: Long term toxicity of a Roundup herbicide and a Roundup-tolerant genetically modified maize. Food Chem Toxicol 50: 4221–4231. doi:https://doi.org/10.1016/j.fct.2012.08.005

81. Steinberg P, van der Voet H, Goedhart PW, et al (2019) Lack of adverse effects in subchronic and chronic toxicity/carcinogenicity studies on the glyphosate-resistant genetically modified maize NK603 in Wistar Han RCC rats. Arch Toxicol 9: 1–45. doi:https://doi.org/10.1007/s00204-019-02400-1

82. Sánchez MA, Parrott WA (2017) Characterization of scientific studies usually cited as evidence of adverse effects of GM food/feed. Plant Biotechnol J 15: 1227–1234. doi:https://doi.org/10.1111/pbi.12798

83. Sleator RD (2016) Synthetic biology: from mainstream to counterculture. Arch Microbiol 198: 711–713. doi:https://doi.org/10.1007/s00203-016-1257-x

84. Grunwald HA, Gantz VM, Poplawski G, et al (2019) Super-Mendelian inheritance mediated by CRISPR–Cas9 in the female mouse germline. Nature 566: 105–109. doi:https://doi.org/10.1038/s41586-019-0875-2

85. Simon S, Otto M, Engelhard M (2018) Synthetic gene drive: between continuity and novelty: Crucial differences between

gene drive and genetically modified organisms require an adapted risk assessment for their use. EMBO Rep 19: e45760–4. doi:https://doi.org/10.15252/embr.201845760

86. Tanaka H, Stone HA, Nelson DR (2017) Spatial gene drives and pushed genetic waves. Proc Natl Acad Sci USA 114: 8452–8457. doi:https://doi.org/10.1073/pnas.1705868114

87. Esvelt KM, Gemmell NJ (2017) Conservation demands safe gene drive. PLoS Biol 15:e2003850. doi:https://doi.org/10.1371/journal.pbio.2003850

88. Reeves RG, Voeneky S, Caetano-Anollés D, et al (2018) Agricultural research, or a new bioweapon system? Science 362: 35–37. doi:https://doi.org/10.1126/science.aat7664

89. Sills J, Simon S, Otto M, Engelhard M (2018) Scan the horizon for unprecedented risks. Science 362:1007–1008. doi:https://doi.org/10.1126/science.aav7568

90. Pearce F (2008) Ozone hole? What ozone hole? New Sci 199: 46–47. doi:https://doi.org/10.1016/S0262-4079(08)62382-9

91. Farman JC, Gardiner BG, Nature JS (1985) Large losses of total ozone in Antarctica reveal seasonal ClOx/NOx interaction. Nature 315: 207–210. doi:https://doi.org/10.1038/315207a0

Further Reading

Eriksson D, Pedersen HB, Chawade A, et al (2018) Scandinavian perspectives on plant gene technology: applications, policies and progress. Physiol Plant 162: 219–238. doi:https://doi.org/10.1111/ppl.12661

Gelinsky E, Hilbeck A (2018) European Court of Justice ruling regarding new genetic engineering methods scientifically justified: a commentary on the biased reporting about the recent ruling. Environ Sci Eur 30: 52. doi:https://doi.org/10.1186/s12302-018-0182-9

Torretta V, Katsoyiannis IA, Viotti P, et al (2018) Critical Review of the Effects of Glyphosate Exposure to the Environment and

Humans through the Food Supply Chain. Sustainability 10: 1–20. doi:https://doi.org/10.3390/su10040950

Upton HF, Cowan T (2015) Genetically Engineered Salmon. Congressional Research Service, Nummer R43518

4

Reading Genetic Material

In 1975, the British biochemist Frederick Sanger published an enzymatic method, and in 1977 the US scientists Allan Maxam and Walter Gilbert published a chemical method for analysing the sequence of nucleotides in DNA, DNA sequencing [1, 2]. Remarkably, Sanger had already developed a method for analysing the amino acid sequence in proteins in the early 1950s and is one of four scientists so far to have been awarded two Nobel Prizes.

Sanger sequencing, also known as the chain-termination method, works on the basis of the enzyme **DNA polymerase**. This is naturally responsible in every cell for duplicating the genetic material before cell division. Enzymes are generally more difficult to handle than chemicals, as they first have to be isolated from organisms and usually do not have a long shelf life. But because it is easier to automate, Sanger sequencing prevailed and remained the dominant sequencing method until the 1990s. However, it also has the disadvantage that sequencing is carried out in two separate steps. In the first step, DNA fragments of different lengths are generated, depending on the DNA sequence to be analysed. This step takes about 2 h. In a second step,

© Springer-Verlag GmbH Germany, part of Springer Nature 2022
R. Wünschiers, *Genes, Genomes and Society*,
https://doi.org/10.1007/978-3-662-64081-4_4

these fragments are separated according to their size either in an agarose gel or in a capillary, which takes several hours. From the pattern of fragment sizes, the sequence of nucleotides of the DNA to be sequenced can finally be determined. Further developments of Sanger sequencing mainly concerned the parallelisation of the analysis. Whereas in the initial phase it was possible to analyse around 10,000 nucleotides per day and device, with modern **capillary sequencers** of the 2000s it was around one million nucleotides.

Sanger sequencing is a **first-generation sequencing method** (Fig. 4.1). It remained the leading method for so long because until recently it allowed the largest read length of approx. 800 nucleotides—nanopore sequencing represents a revolution in this context (see later).

800 nucleotides? The human genome contains 3.2 billion nucleotides; how can that be? Let us think of the genome as a book. With Sanger sequencing, the first 800 characters can be read in about 3 h. In another 3 h, the next 800 characters, and so on. To be faster, we cut the book into many snippets of 800 characters each and can read them in parallel in 3 h. But in the end, how do you know which

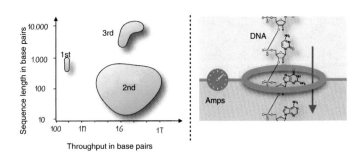

Fig. 4.1 (Left) Comparison of the performance of different sequencing technologies of the first, second (*next generation sequencing*) and third generation. M = Mega, G = Giga, T = Tera. (Right) In nanopore sequencing, the DNA molecule is passed through a pore. The sequence can be derived from the resulting current fluctuations

strings are consecutive? You don't! So, you need at least two books, each snipped at different points. For example, one book starting at the first character and then every 800 characters and another book starting at character 200 and then every 800 characters. Then we get overlapping snippets and can reconstruct the sequence of characters in the original book. This is how genome sequencing works (Fig. 4.2). The many sequence fragments of 800 nucleotides each (reads) must be assembled to form the whole genome. This *assembly* is an important step and works better the longer the sequence analysed and the more genome copies are sequenced. Thus, when it is said that a genome has been sequenced, in reality thousands of copies have usually been analysed (Fig. 4.2).

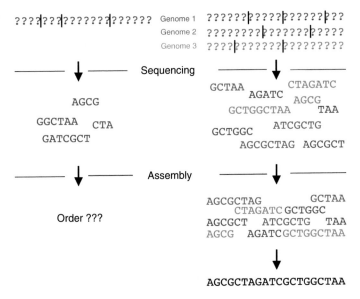

Fig. 4.2 Schematic sequence of a genome sequencing. (Left) With only one DNA molecule as a template, no sorting of the fragments would be possible. (Right) With several copies of the DNA to be analysed, overlapping fragments arise. Therefore, the DNA sequence can be assembled correctly

In 1996, the Swedish biochemist Pål Nyrén from the *Royal Institute of Technology* in Stockholm, together with his PhD student Mostafa Ronaghi, developed the so-called **pyrosequencing** method [3]. It has nothing to do with fireworks, but much more to do with detecting the molecule pyrophosphate (also called diphosphate). The pyrophosphate is released when the DNA polymerase—as in Sanger sequencing, again the basis of sequence analysis—is active. A light signal is then released via several enzymatic steps. The revolutionary aspect of this **second-generation sequencing technology** was the combination of molecular reading and result output [4]. Thus, with pyrosequencing, the result is obtained almost in real time. In addition, the technical analysis system could be made considerably smaller, so that many more reactions can be analysed in parallel per device. But: The length of the sequences (reads) is reduced to about 100 nucleotides. It depends on the scientific question whether this length is sufficient.

Lasting history in combination with pyrosequencing technology was made by the Swedish **company 454** in 2007, when they used their *Genome Sequencer FLX sequencer* to decode the complete genome of the co-discoverer of DNA structure and Nobel Prize winner James Watson [5]. Although the project was certainly intended as a publicity stunt for their technology, *454* (then already part of *Hoffmann-La Roche*) showed that within 4 months, with a handful of scientists and a little less than US$1.5 million, a complete human genome could be deciphered. When the results of **Project Jim** were presented to the public, Watson made only one condition: He absolutely did not want information about his alleles of the *apoE* gene to be disclosed (it encodes apolipoprotein E, which plays an important role in fat metabolism). This is because variant four of this gene is associated with an early onset of Alzheimer's—Watson was 79 years old at the time of his sequencing. It is fair to say that Project Jim was the first major sequencing project of a new generation of high-throughput sequencers—and the second fully sequenced human genome. For also in 2007, the US biochemist Craig Venter presented his genome sequence [6]. This sequencing was carried out using the Sanger method on high-throughput sequencers called ABI 3730xl from the company *Applied Biosystems.* The project took about 7 years and cost about US$100 million. Based on his genetic data, Venter also published his **autobiography** in 2007—the first to relate life events to gene variants [7].

Both projects took advantage of the fact that a first draft of an incomplete sequence of the human genome was published in 2000. This data helped both teams to piece together the *reads.* Today, thousands of individual human genomes have been deciphered, even that of **Neanderthal man**, and the cost per genome is less than US$1000. Just recently, I received a combined *Black Friday-Cyber Monday* offer from US-based *Dante Labs.* They offered a complete genome sequencing for €169 (Fig. 4.3).

Fig. 4.3 A complete genome sequencing special offer from the US company *Dante Labs.* Screenshot of the author

In the 2000s, numerous new sequencing methods were developed and brought to market. Automated sequencers became more compact, cheaper and able to analyse more nucleotides per time and device. This was mainly due to advances in microfluidics. Modern automatic sequencers are about the size of a laser printer, cost around 40,000 euros and can analyse 100 billion nucleotides per day.

A new era of genome analysis was heralded in 2013 by the English company *Oxford Nanopore* when, during a conference in Marco Island, Florida, it presented an automated sequencer that fits in the palm of your hand and can be connected to the USB port of a commercially available computer: the MinION. This **third-generation sequencing machine**, on the market since 2014, is based on **nanopores** in a membrane through which the DNA molecule is pulled by an electric field [8]. This is associated with a change in the current flow through the pore, which in turn is measured (Fig. 4.1). What is revolutionary is, on the one hand, the compactness of the device (Fig. 4.4) and, on the other hand, the fact that sequences of several thousand nucleotides can be read at a time.

Mobile sequencing with nanopores will not only revolutionise genetic diagnostics [9]. **RNA** can already be

Fig. 4.4 Highly delighted, the author receives one of the first MinION deliveries in Germany in July 2016. The mobile sequencing machine is connected to the computer via the USB port. (Photos: R. Leidenfrost and R. Wünschiers)

analysed using this method, and in the future, it will certainly also be possible to analyse **proteins** [10]. Nanopore technology could then be used to identify individual proteins or protein modifications (post-translational modifications) and analyse DNA-protein interactions. The fact that the MinION is a handheld device means that samples to be analysed no longer have to be sent to a service provider, but can be analysed in real time on site (at the patient's bedside, in the field, etc.). It is very conceivable that in the near future small sequencing machines will be sold in educational kits for children in toy stores. A prospect that will occupy us in Sect. 7.1.

4.1 Genetic Variation in Humans

Our genome is 3.2 billion nucleotides long (Sect. 2.1). My genome is at least 99.5% identical to yours, 99% identical to that of a chimpanzee and 45% identical to the genome of a lettuce. What are the genetic differences, also called **polymorphisms** (Greek polymorphism)? Numerous international projects are devoted to this topic [11]. For example, in October 2015, a study was published that examined genetic variation in 2504 people [12]. A 2018 study already examined 17,795 individuals [13]. There are small differences at the sequence level that can only be studied by sequencing the genome or similar methods. And there are large differences, some of which can even be seen under the microscope (Fig. 4.5). In total, the polymorphisms consist of about 60,000 longer insertions and 3.6 million shorter insertions or deletions. With a number of around 85 million, *single nucleotide polymorphisms* (**SNPs**) account for the largest proportion of genetic variation in humans (Sect. 4.2). Two people differ in an average of 16 million base

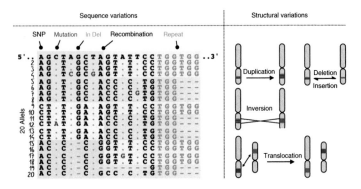

Fig. 4.5 Variants (polymorphisms) in the genome at the level of DNA sequences or chromosomes. (Left) Twenty variants (alleles) of a gene are shown. Nucleotides that are identical to the top sequence are shown as grey dots. Frequently occurring variations are single nucleotide polymorphisms (SNPs). Rare deviations are mutations (red). Deletions are shown as blue dashes. TGG repeats are marked in green. At recombination sites, exchanges of large DNA fragments often occur, as in translocation. (Right) Representation of larger scale structural polymorphisms at the chromosome level. The changes can occur either within a chromosome (intra-) or between chromosomes (interchromosomal)

pairs. They determine, for example, eye colour or blood group. Likewise, these genetic differences, individually or in combination, determine the risk of disease outbreak (Sect. 4.2). Around 80 specially selected SNPs, distributed over the entire genome, are sufficient to uniquely identify a person [14]. Even identical **twins** can be distinguished in this way [15].

The causes of polymorphisms are diverse and elementary. Diverse because they are caused both by errors of the molecular machinery in the creation of a copy of the genetic material (**replication**) before cell division, and by external influences such as radiation or chemicals. Structural variations in chromosomes occur primarily during **meiosis**. This is a process involved in the formation of germ cells, that is eggs and sperm. Meiosis is therefore reserved for organisms with sexual reproduction. The mechanism by which most structural variants arise is called *crossing-over* (Fig. 2.6). A variation that initially emerges in an individual is called a mutation. However, if this mutation is passed on to off-spring and thus spreads over generations in a group of individuals (population), then we call it polymorphism. Strictly speaking, **polymorphism** must occur in at least 1% of all alleles. In the case of a single nucleotide **mutation**, we call the polymorphism a **SNP** (*single nucleotide polymorphism*). We will see later that it is these SNPs that play an important role in genetic diagnostics.

Variations are crucial because they are the engine of evolution. Variants can have different strengths of activities or even functions. In this way, selection can choose the best adapted individuals to given conditions. They have the greatest success in reproduction, have the greatest fitness. This, by the way, is one of the great problems for everything that takes place after the reproductive phase: selection pressure decreases. We pass on half of our genetic material to our children and a quarter to our grandchildren. If living beings no longer actively participate in the preservation of the hereditary material that is passed on, whether through care, nutrition or, for example, the passing on of culture, then the processes of evolution have at best a neutral effect. There is probably no selection pressure against, for example, cancer in old age. The saying of the Russian-American evolutionary biologist Theodosius Dobzhansky still applies:

> Nothing in biology makes sense except in the light of evolution [16].

It is important to understand that there is no *one gene*. If a gene is responsible for a disease, then there are two variants of this gene in a single person, the **alleles**. If these two variants are exactly identical, then we say that our genome is **homozygous** with respect to that gene, or monomorphic. If we have two different copies on the maternal and paternal chromosomes, then we are **heterozygous** or dimorphic. If we now look at two individuals, there are four alleles of a gene. These can all be identical, but they can also be different. In the Fig. 4.5 for example, twenty alleles are shown. Alleles 1 to 3 and 5, 6, 8 and 9 to 10 and 11, 13, 14 and 15, 16, 18 are identical. So, there are eleven different alleles, of which each individual can carry two variants. This yields already 121 possible combinations or more precisely, **genotypes**. As already described, alleles can differ in their biological activity. Different combinations of the two alleles in a genome can therefore lead to different manifestations, to **phenotypes** (Sect. 2.1). For our example from just now, this would be 121 theoretically possible phenotypes. Most of the characteristics that make us who we are, are therefore not based on differences in individual genes but are determined by the combination of alleles present. However, the effect of a **dominant** allele can outshine the effect of a **recessive** allele. This is shown in Fig. 4.6.

I had written that our two genomes, yours and mine, differ by no more than 0.5%. That is a maximum of 16

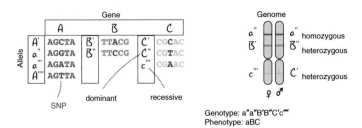

Fig. 4.6 There are between two and four alleles of three genes (A–C). Capital letters stand for dominant, small letters for recessive alleles. In the genome, one maternal and one paternal allele come together. Recessive alleles must be present on both chromosomes (homozygous) in order to express in the phenotype

million different positions, some of which are SNPs. But what does this look like between people from Africa, Europe or Asia, for example? Such questions can be asked nowadays because there are enough sequenced humans, for example in the framework of the *1000 Genomes Project*. In 2014, for example, the Swedish evolutionary geneticist and director at the *Max Planck Institute for Evolutionary Anthropology* in Leipzig Svante Pääbo (Sect. 4.2) examined the genetic material of 185 Africans and 184 Europeans and Chinese (Eurasians) [17]. He found 38,877,749 positions where the genomes differed. However, he could not prove a single position where all Africans differed from all Eurasians. Thus, you could not develop a genetic test that could distinguish African people from European or Chinese people 100% of the time. And at 95%? There are twelve positions. This means that for only twelve of the 3.2 billion nucleotides we cannot be 100% sure, but we can be 95% sure that the genome, or the person, comes from Africa. Further, someone in our neighbourhood who has lived next door for generations could be more genetically distant from us than a person who has only recently moved from Africa. Why is that?

Let us take a look at the Fig. 4.7. The circles schematically represent the genetic variance within a population of

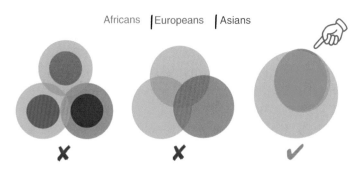

Fig. 4.7 Original and actual models of the genetic variance of the human genome

African, European or Asian people. In the left-hand model, we assume that there are polymorphisms that occur only in Africa, Europe and Asia (dark core) and genetic variants that are shared by the respective populations. But this model is wrong. The middle model is similar. But it does not assume that there are population-specific variants. Instead, there is a statistically high proportion that is population-specific and a proportion that is common to the populations. But this model is wrong, too. On the contrary, it has been shown that African populations have a much greater genetic variance than Europeans and Asians.

Although there are about six times as many people living outside Africa as inside it, the existing genetic variation within Africa is much greater. This is because modern humans evolved in Africa and thus had more time to accumulate genetic diversity there (Sect. 4.2). Then, a small subpopulation, a **founder population**, probably left Africa 100,000 years ago and spread to Europe (*out-of-Africa theory*). Thus, only a small part of the polymorphisms left Africa. Only since then have they been able to develop new, "own" polymorphisms or, for example, "collect" them from the Neanderthals or Denisovans (Sect. 4.2) by mating (see hand in Fig. 4.7). According to this model, we are basically all Africans. The only thing that distinguishes us from them is our kinship to Neanderthals and, in the case of the inhabitants of Australia and Oceania, additional kinship to Denisovans [18].

These findings from genome analyses also show once again that the concept of **race** makes genetically no sense. Different populations differ only to a negligible and statistically only weakly significant extent in the presence or absence of polymorphisms in alleles. Rather, they differ in their distribution (Fig. 4.8).

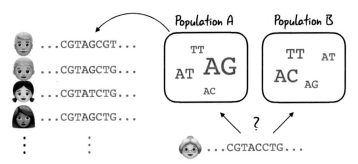

Fig. 4.8 Your neighbour could be genetically more distant from you than an immigrant from distant latitudes. Polymorphisms such as SNPs differ less in their presence or absence than in their frequency in a population

4.2 Genetic Diagnostics

A quantum leap in the analysis of genetic information was achieved with the development of the polymerase chain reaction (**PCR**) by the US biochemist Kary Mullis in 1980. As in Sanger sequencing, the enzyme DNA polymerase is used. This method makes it possible to amplify minute quantities of DNA molecules and thus make them accessible for genetic analyses. The available methods are manifold. They range from the detection of length differences of specific regions as a result of nucleotide insertions or deletions due to natural mutations (they are the basis for the **genetic fingerprint** in forensics or paternity testing) to the detection of the exchange of individual, specific nucleotides (**SNPs,** *single nucleotide polymorphisms*) by sequencing the amplified DNA sections.

The possibility of multiplying the genetic material also means that the smallest amounts, for example skin abrasion on a table, are sufficient to obtain enough DNA for genetic analysis. This pleases the honest person in the case of clearing up offences. These can also be of an unusual nature: recently, for example, genetic analyses from confiscated elephant tusks were compared with geographical data sets, thus revealing trade routes [19]. However, the methodology also raises the question of **genetic data security.** This issue is becoming increasingly topical due to the drop in prices and the simplification of DNA sequencing (see earlier in this chapter).

The insights we can gain from the genetic information of long-dead organisms are fascinating. Svante Pääbo (Sect. 4.1) is a pioneer in the field of **palaeogenetics**. He developed the methodological tools to isolate and sequence DNA from ancient biological material such as mummies or bone finds. In 2010, palaeogenetics gained great attention with the publication of a large part of the DNA sequence of a **Neanderthal man** at least 35,000 years old and a **Denisova man** at least 30,000 years old (Fig. 4.9) [20, 21].

The results shed new light on the evolutionary and migration history of modern humans [18]. Around four to

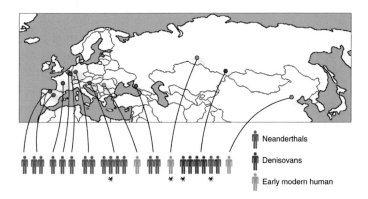

Fig. 4.9 Locations of our ancestors. Only finds of which a part of the genome could be examined are shown. The genetic material of the finds marked with an asterisk were completely sequenced. Highlighted is the find of a female offspring of a Denisovan father and a Neanderthal mother (red/blue). (Source: Slon et al. [22])

7% Neanderthal or Denisovan DNA in our genome tells its own story of more or less romantic evenings between *Homo sapiens* and *Homo neanderthalensis* around the campfire. Only recently, the genome of a girl who died over 50,000 years ago was analysed, revealing that her mother was a Neanderthal and her father a Denisovan (Fig. 4.9) [22]. Both Denisovans and Neanderthals were replaced by our direct ancestors about 40,000 years ago.

The genome of the *Iceman*, better known as **Ötzi**, discovered in the **Ötztal Alps** in 1991, was also decoded and presented to the public in 2012 [23, 24]. The genetic material from the Copper Age, which is at least 5300 years old, tells us, for example, something about the geographical origin of his ancestors (Corsica), his eye colour (brown), his blood group (0), a Lyme infection he experienced, his lactose intolerance and a predisposition to the disease of his coronary vessels [25]. How is all this possible?

A decisive contribution in terms of diagnostics is made by so-called **association studies**, that investigates which genetic information is associated with which clinical picture [26]. It can be very simple: a mutation in a gene is responsible for a disease (Fig. 4.10).

Currently, more than 2700 such **monogenic** diseases are known. Nevertheless, the situation is usually more difficult, since there are several genetic variants (**alleles**) of such genes. **Cystic fibrosis** (or mucoviscidosis, a disease characterized by viscous body secretions), for example, is a monogenic disease that affects one in 2000 new-borns in Germany. However, there are more than 1000 known variants of the gene responsible, called *CFTR* (*cystic fibrosis transmembrane conductance regulator*). It is about 250,000 nucleotides long and codes for a protein of 1480 amino acids. There is a lot of room for mutations. However, most diseases are **polygenic** and are therefore caused or

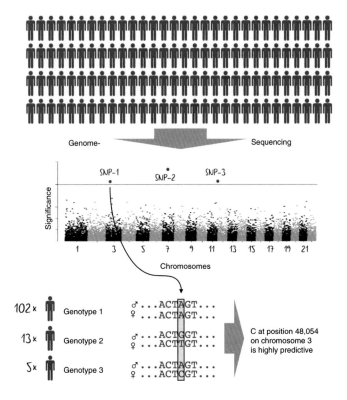

Fig. 4.10 Genome-wide association studies help to develop genetic tests for diseases. Here, all 22 chromosomes (excluding sex chromosomes) were sequenced from 120 individuals, five of whom have a disease (red). The diagram shows for 16,427 variable positions (SNPs) in the genome the likelihood that they are to occur in people with the disease. For three SNPs (1–3), the significance is very high. For one of these SNPs (SNP-1) the three genetic combinations (genotypes) found are shown. A cytosine (C) at position 48,054 on chromosome 3 occurs only in diseased individuals and may form the basis for a genetic diagnostic test or even therapy

influenced in their expression by several genes. The type of mutation in the genes or alleles can also be diverse and can range from duplications to insertions or deletions of nucleotides to simple nucleotide exchanges, the so-called SNPs. For association studies, a fixed set of about 720,000 SNPs is usually examined. This reduction of the genome from 3.2 billion to 720,000 nucleotides is roughly analogous to simplifying a molecule to its atoms—with the same effect: it helps to develop models. However, one cannot necessarily infer the molecule from knowledge of the atoms present. For example, many SNPs are known to be related to a disease but are not directly located in a coding gene. Such SNPs therefore have diagnostic value, but cannot provide direct approaches for therapy because the mode of action is unknown. They usually do not lead to an altered structure of a protein or enzyme, but to a change in gene activity. Since DNA lies like a ball in the cell nucleus, even distant areas can touch and influence each other [26].

Of the 3.2 billion nucleotides in the human genome, over 320,000 positions that have been clinically tested are known to be associated with disease. It would therefore be sufficient to analyse these 320,000 positions in the genome to be able to make statements about the state of health and the risk of disease. In the age of self-measurement and self-optimisation, this naturally generates a market (Sects. 7.2 and 7.3).

The **Gene Diagnostics Act** came into force in 2010 and regulates genetic examinations for medical purposes, to clarify parentage, as well as in the insurance sector and in working life. The Act regulates the requirements for genetic analyses and the use of the data obtained from them. This is intended to,

> to prevent discrimination on the basis of genetic characteristics [... and ...] to uphold the State's obligation to respect and protect human dignity and the right to informational self-determination.

The law distinguishes between diagnostic and predictive genetic tests. A **diagnostic examination** serves to clarify an existing disease or health disorder and to determine genetic characteristics which, together with other influences (e.g., interaction with medicines), may trigger the onset of a disease or disorder. Such an examination may only be carried out by doctors and the results may only be disclosed by them. A **predictive examination** serves to clarify a disease or health disorder that may occur in the future and to clarify the existence of a genetic predisposition that could be passed on to offspring. Predictive genetic testing may only be carried out by specialists in human genetics or by those who are specially qualified in human genetic testing. In addition, the patient must consent to the genetic examination and be fully informed beforehand, including about the significance for future life. The major difference between the currently prevailing diagnostics, which focus on specific measured values, and genome analysis (and also transcriptome analysis; see below) is the density of information generated about a patient. Thus, a genome-wide analysis does not represent a punctual, but rather an ongoing encroachment on **information rights**, since more and more information can be obtained gradually from the stored genetic data. This applies regardless of whether the sequencing had a diagnostic (acute) or a predictive reason. Here it is important to clarify that patients may learn things about their genetic make-up that are more unsettling than helpful.

Whereas individual genes, sections of genes or entire genomes have been analysed in diagnostics up to now, the latest trend is towards RNA. **RNA diagnostics** (or transcriptome diagnostics, also known as RNA-Seq) is used to examine which genes are active. It is therefore the living part of the genome. This form of diagnostics can be performed for individual genes, but also for the entire genome. While the information of a single gene read by the cell is called a transcript, the totality of all transcripts in a cell is called a transcriptome, analogous to the term genome for the totality of all genes. Two studies published in 2017 demonstrated that the method of transcriptome sequencing could diagnose diseases that were not identified by standard genetic diagnostics [27, 28]. In addition, new candidate genes were found for the respective diseases. With the help of such additional information, a diagnostic test can be improved step by step as the number of tests increases.

4.3 Prenatal Diagnostics

Since the early 1970s, parents of new-borns in Germany have had the right to have their offspring tested for rare, incurable, but severe congenital, metabolic, and hormonal diseases. The current **new-born screening** is usually carried out on the third day of life in conjunction with the U2 (first basic medical examination after birth) and covers 15 genetic diseases. Participation in the diagnostic procedure is voluntary and the costs are covered by the statutory health insurance funds. The purpose of the examination is to be able to start treatment for incurable diseases as early as possible. With a screening before birth, other options for action are available. This form of diagnostics has developed rapidly in recent years, not least because of the increase in artificial insemination. Before we approach the subject, I would like to recall the development of the fertilized egg into an embryo (Fig. 4.11).

Artificial insemination (***in-vitro* fertilisation**, IVF; *assisted reproduction technology*, ART) is basically the creation of a pregnancy without mating, that is sexual intercourse. The development of new reproductive technologies began with animal breeding. It has nothing to do with genetic engineering in the narrower sense, but to a certain extent it

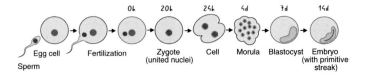

Fig. 4.11 The natural course of embryogenesis from fertilisation to the formation of the embryo in the second week. In *in-vitro* fertilisation, the blastocyst is transferred to the uterus after 2–6 days [29].

paves the way for it. The fact that the fertilisation of the egg cell by the sperm is moving further and further away from humans, starting with the injection of sperm into the genital tract of the woman (**insemination**) and ending with artificial insemination, creates a lot of space for the application of techniques. In addition, emotional distance is created. Sexual intercourse in our society is already largely detached from a reproductive goal anyway. Rather the opposite, the potential bringing about of conception is avoided by a wide variety of contraceptive methods. And just as mechanisms to control not-getting-pregnant are increasingly available, so are methods to control getting pregnant. The US geneticist Henry T. Greely believes that methods of reproductive medicine, which currently mainly benefit infertile couples, are becoming the norm. Accordingly, he titled his 2016 book *The End of Sex and the Future of Human Reproduction* [30]. In it, he cites two factors that he believes will lead to a steady increase in *in-vitro* fertilizations, regardless of infertility. First, new techniques are becoming easier to use and also therefore cheaper, and second, economic, social, legal, and political forces are pushing. At the centre of his argument, which is shared by many experts, is what Greely calls ***Easy PGD*** (*preimplantation genetic diagnosis*). This simple pre-implantation genetic diagnosis combines the advances of diagnostics with those of stem cell research. But more on that later. In addition, there is an aspect that repeatedly leads to discussions in various areas of medical care: **overtreatment** (unnecessary health care). This means that more *in-vitro* fertilisations are carried out than seems necessary. Studies have shown, for example, that although one in seven couples try for more than 12 months to fulfil their desire to have a child, after a further 12 months only about half of them become successful parents—without IVF [31]. In contrast, the success rate of *in-vitro* fertilisation is around 30% per treatment carried out.

At this point I would like to draw attention to a problem with **insemination**. I may be straying from the subject for a moment, but it shows the importance of the human factor, which is not insignificant in terms of risk assessment. Insemination, in which sperm is injected into a woman's genital tract, was first used probably in **dog breeding.** The English physician and dog breeder John Everett Millais describes in his 1889 published book *The theory and practice of rational breeding* the implementation of the theory of evolution according to Charles Darwin and Thomas Henry Huxley as well as the theory of heredity of Francis Galton, all contemporaries of Millais, for dog breeding [32]. As he published 1884, insemination contributes to an increased mating success, securing of the pedigree and avoidance of the transmission of diseases [33, 34]. From the 1930s onwards, insemination was also used in cattle breeding, where it has become indispensable. Insemination centres usually supply several thousand cattle farmers with, for example, bovine semen for around €40 per millilitre. Until the 1960s, the use of insemination in humans was banned in many countries and was first legalised in the USA in 1964 in the state of Georgia. Insemination clinics in the livestock industry were joined by **fertility centres**. The US-American reproductive physician Cecil B. Jacobson was the first to introduce *amniocentesis* for diagnostic purposes in the USA in 1967 [35]. However, he became famous for his **unethical behaviour** as a reproductive physician. This provided the material for a US television film entitled *The Babymaker: The Dr. Cecil Jacobson Story*, also marketed as *Seeds of Deception* from 1994 and the documentary *The Sperminator* from 2005. It also led to a five-year prison sentence in 1992: Jacobson not only deceived women with false pregnancies brought about by hormone injections, he also used his **own sperm** for artificial insemination. But Jacobson was not the only known case. In December 2017, then 79-year-old US physician Dr. Donald Cline was sentenced to a one-year suspended prison term for the same misconduct. Of more than 35 donor offsprings, as they call themselves, Cline is known to be the biological father. His misdemeanours occurred from the 1970s to the 1980s, while he was working at a fertility clinic in Indianapolis. At that time, he had no way of knowing that modern parentage analysis would help uncover the cases.

In fact, it is in particular the providers of parentage testing such as *23andMe, deCODE Genetics* or *Navigenics,* which are popular in the USA, who nowadays bring together half-siblings, who then often carry out further research (Sect. 7.1 and Fig. 7.3). In the meantime, there are separate initiatives with corresponding databases and websites. The Dutch reproductive physician Dr. Jan Karbaat, who died in April 2017 at the age of 89, fathered at least 19 donor offsprings. Since the children of his patients ranged in age from 8 to 60, the number of unreported cases is probably much higher. He himself insisted until his death that he had never used his own sperm for the treatments. An initial suspicion arose when DNA analysis of one of his biological children led to numerous half-siblings—some 200 direct descendants of Karbaat are now believed to exist.

By law, no more than six children are allowed to be conceived from donor sperm in the Netherlands, and also in Germany. Some descendants organized themselves and in 2017 filed a lawsuit to be allowed to examine their biological parentage. In February 2019, a court in The Hague allowed them to compare their DNA sequences with those of the deceased doctor. This was isolated from personal items such as his toothbrush after his death in 2017.

In Germany, the Federal Constitutional Court ruled in 1989 that the general right of personality enshrined in the Basic Law also includes the **right to know one's own parentage**. Sperm donors thus have no right to anonymity, even if no maintenance or inheritance claims are associated with the annulment. In July 2017, the *Act regulating the right to know one's parentage in the case of heterologous use of semen* (the so-called Semen Donor Registry Act, *Samenspender-Registergesetz*, SaRegG) was additionally passed. Since July 2018, the German *Federal Ministry of Health* has maintained a nationwide sperm **donor register**. Here, personal data of sperm donors and recipients are stored for 110 years.

There are numerous other cases, in Germany for example the case of Professor Thomas Katzorke. As parentage analysis has become easily accessible and cheap, more cases of ethical boundary crossing in the doctor-patient relationship are sure to follow. In all cases, the doctors deceived their patients, as they were led to believe that they had been injected with anonymous donor sperm. It is certainly not an exaggeration to speak of rape here. What do these cases teach us? Probably, above all, that reproductive physicians have a great responsibility that can be comparatively easily abused. But for all the ethics and all the personal suffering of those involved, there is still a population biology aspect: the medical professionals have overrepresented themselves and gained a place in the **gene pool** (Chap. 2). This may have long-term consequences, as known from inbreeding: The accumulation of rare, pathogenic alleles (Sect. 2.1) and thus an increased risk of hereditary diseases. How fatal the consequences can be is shown by the close kinship network in hereditary dynasties such as the Habsburgs with **cousin marriages** or marriages between uncles and nieces: within a few generations many children died before they themselves could give birth to offspring [36, 37]. In the case of fourth-degree blood relatives (first cousins), this risk is doubled (Fig. 4.12).

While this purifying selection (or extinction) of deleterious alleles is beneficial to the overall population, it could be avoided by more distant marriages. Interestingly, a study of

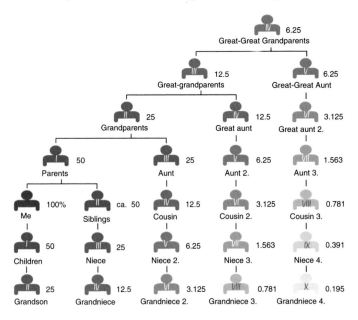

Fig. 4.12 Kinship relations and designations in the family tree. Roman numerals indicate the legal degree of kinship. The numbers next to the persons indicate in percent the match of the hereditary property (kinship coefficient)

all known Icelandic couples and their descendants between 1800 and 1965 found that eighth- and tenth-degree consanguineous marriages (third and fourth cousins) had the greatest reproductive success [38]. Clearly, there is an optimum for genetic distance (**degree of relatedness**) in mating in terms of the disadvantage of increased risk for inherited diseases and the advantage of preserving well-matched (co-adapted) gene complexes [39]. What these gene complexes may be is still completely unknown. For the analysis, 160,811 Icelandic couples were studied. However, the inhabitants of Iceland, which was settled by Vikings and Celts about 1100 years ago, are a comparatively homogeneous island population that has had little genetic exchange with other populations for decades [40]. But even if the findings of a biologically optimal kinship ratio could be transferred, the maximum number of children was never 40 or upwards.

Why this digression? In my view, the introduction of the insemination method to fulfil a desire for children opened Pandora's box. It was the first step in the mechanisation of human reproduction, which the writer and Georg Büchner Prize winner Sibylle Lewitschroff described rather unhappily during her 2014 Dresden speech entitled *From Feasibility. The Scientific Determination on Birth and Death* rather unhappily referred to as *reproductive mess*. The aforementioned doctors were probably the first to abuse the techniques out of selfishness—presumably they were not at all concerned with enriching the gene pool with their alleles, but rather with financial and perhaps emotional self-enrichment. But we must always bear in mind that the **power of reproductive medicine** is great, because it has an effect on subsequent generations. This applies all the more to the selection or genetic modification of embryos (Chap. 5). The latter has ushered in a new era with the highly controversial gene editing by the Chinese biophysicist He

Jiankui at the end of 2018 (Sect. 5.1). Retrievability, as demanded by critics of the release of genetically modified organisms, for example, is unthinkable for ethical reasons.

The distance of the fertilization process from the body grows during ***in-vitro* fertilization** (IVF). *In-vitro* stands for in a glass, as in vitrine. Louise Joy Brown was born on July 25, 1978, and was the first person conceived using *in-vitro* fertilization. The first German child conceived with the help of IVF was born on April 16, 1982. The number of *in-vitro* fertilizations in Germany has multiplied from 742 treatments in 1982 to around 90,000 completed treatments in 2016. Worldwide, over five million children like Joy Brown have already been conceived through IVF. In well over half of all cases, this involves injecting the male sperm into the egg using a type of syringe (Fig. 4.13) and in around a quarter the sperm had previously been frozen. The latter is increasingly being offered to female employees by companies such as *Apple* or *Facebook*: Freezing of their egg cells in order to pursue the desire to have children in the future, after a successful career and establishment in the company (or better another?). This procedure is called

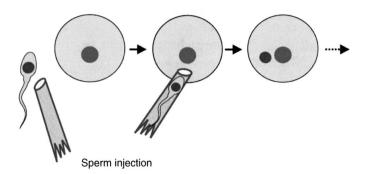

Sperm injection

Fig. 4.13 A special form of *in-vitro* fertilisation (IVF) in intracytoplasmic sperm injection (ICSI), in which the male sperm is injected directly into the female egg cell

social freezing and is not forbidden in Germany, unlike egg donation. For women who want to have children after menopause, it is a legal way to conceive a child. In 2015, a 65-year-old retired Berlin woman gave birth to quadruplets after artificial insemination. The number four alone shows that the implantation of the fertilized eggs did not take place in Germany, where the number is limited to a maximum of three embryos, but in this case in Ukraine. And that would not have been possible in Germany either: both the sperm and the egg cell came from donors.

But if the fertilisation process has to be moved out of the body and into a laboratory for medical reasons, then the barrier to accepting further medical services is lowered. After all, if one already has to endure the physically and often emotionally stressful procedure of egg retrieval, then one also wants to maximize reproductive success. The decision about an abortion could also be made early on with a genetic test—in extreme cases even before fertilisation. Already today, companies send sperm from donors who are specifically selected. This goes as far as listening to the voice or looking at the handwriting. In September 2013, scientists from the company *23andMe* were granted US patent number 8,543,339 for the *selection of germ cells based on genetic calculations* [41].

In the field of prenatal diagnostics of embryos and foetuses, a distinction must be made between analysis during pregnancy (**prenatal diagnostics**) and examination within the framework of reproductive medicine, that is before a possible transfer of an embryo into the uterus (pre-implantation diagnostics). In Germany, prenatal diagnostics, like the genetic examination of born human beings, is governed by the **Gene Diagnostics Act**, which stipulates that a genetic analysis may be carried out only for medical purposes and only if the examination is aimed at certain genetic characteristics of the embryo or foetus which, according to the generally recognised state of scientific and technical knowledge, affect its health during pregnancy or after birth, or if treatment of the embryo or foetus with a medicinal product is planned, the effect of which is influenced by certain genetic characteristics.

The pregnant woman, in turn, must be informed and consent to the examination. On the other hand, an examination in relation to a disease which, according to the recognised state of scientific knowledge, can only break out after the completion of the 18th year of life, is not permitted. A special, non-invasive form of prenatal diagnostics is the **prenatal paternity test**. It can be carried out from the ninth week of pregnancy. The procedure makes use of the fact that DNA of the embryo can be isolated from the maternal blood and analysed. **Abortion** is legally regulated in Germany and is possible up to the twelfth week. Up to the 22nd week, a pregnant woman remains exempt from punishment in the event of an abortion, but not the attending physician. Special regulations apply to later points in time.

Prenatal diagnostics are already being used to reduce the incidence of certain hereditary diseases [42]. **Thalassemia**, for example, which is also called Mediterranean anaemia and is related to sickle cell anaemia, is caused by a genetic defect due to which too little of the red blood pigment is produced (Fig. 2.3). Ultimately, this causes an undersupply of oxygen to the affected person. However, the defect also protects against malaria infections, which is why the disease is widespread in the Mediterranean region. Around 5% of the world's population are carriers of a defective gene. There are several forms of thalassemia, of which β-thalassemia is the most common. Since the 1970s, diagnostic programs have been in place to curb the spread of the defective genes and thus the disease. Currently, there are awareness campaigns and prenatal as well as preimplantation diagnosis programs especially in Italy, Greece, Cyprus, England, France, Iran, Thailand, Australia, Singapore, Taiwan, Hong Kong and Cuba. In the Netherlands, Belgium and Germany, there is at least one specific pregnancy counselling service. In Cyprus, the Orthodox Church insisted on genetic testing before marriage until 2004. In Lebanon, Iran Saudi Arabia, Tunisia, the United Arab Emirates, Bahrain, Qatar, and the Gaza Strip this still applies to Muslims. In Cyprus, the strict control was officially justified with an imminent collapse of the health system, since the number of sick people increased on the one hand and on the other hand the ill people became older and older due to the therapy possibilities. Through counselling, screening, abortions and targeted *in-vitro* fertilisation, the number of new cases has been reduced from 70 per year in the mid-1970s to around two per year today. In Sardinia, too, the number of new-borns with the disease has been reduced to about one tenth.

Another severe disease is **cystic fibrosis**, also called mucoviscidosis (Sect. 4.2 and Fig. 2.3). Affected persons cough up thick mucus and suffer from pneumonia and shortness of breath. Often, diabetes is an additional factor. The disease is treatable but not curable. The Federal Joint Committee, the highest body in the German healthcare system, has been recommending prenatal diagnostics since 2016.

The justifiability of the use of prenatal genetic diagnostics is all the easier, the greater the risk of the occurrence of genetic diseases. The greatest risk is posed by recessive alleles (Sect. 2.3 and Fig. 4.6), which generally do not cause symptoms in carriers of only one copy (heterozygous situation). However, if both the maternal and paternal alleles are defective (homozygous situation), disease occurs. The risk therefore becomes high when a healthy couple wishes to have children, but both partners carry a defective allele.

This risk is increased, for example, in genetically isolated populations and is known from inbreeding (see above). Central and Eastern European Jews, so-called **Ashkenazi Jews**, and their descendants are believed, based on genetic studies, to be descended from a group of about 350 people who lived about 700 years ago [43]. It was and is common for them to **intermarry** (**endogamy**), which prevents the entry of new, diverse alleles. Genetic variability is therefore much lower among them than, for example, among Icelanders or population groups such as the Amish or the Hutterite Brethren. The latter two are also endogamous religious communities. Ashkenazi Jews today make up about 70% of all Jews and are thus 30th cousins. As a result, numerous genetically caused diseases such as Parkinson's disease, breast and ovarian cancer, and Tay-Sachs syndrome are disproportionately elevated in Ashkenazi Jews compared to other population groups. Since the 1970s, genetic tests have been carried out, which are fully accepted and have led to a decrease in the number of diseases.

In contrast to prenatal diagnostics, **pre-implantation diagnostics** (PID) is not regulated in Germany in the Genetic Diagnostics Act. In 2011, however, a regulation was made via the Embryo Protection Act (*Embryonenschutzgesetz*). In it, pre-implantation diagnostics is punishable and is only permitted in special exceptional cases. On the one hand, this concerns cases in which there is a high risk of a serious hereditary disease for offspring due to the genetic predisposition of one or both parents. In the view of the legislator, a serious hereditary disease exists if the disease develops

> significantly different from other hereditary diseases due to a low life expectancy or severity of the clinical picture and poor treatability.

PID is banned in Italy and not regulated at all in Ireland and Luxembourg. In Austria, Switzerland, Belgium,

Denmark, France, the Netherlands, Norway, Portugal, Sweden, Spain, the United Kingdom and the United States, it may be performed under certain legally regulated conditions.

A genetic examination of the germ cells before fertilisation is currently only possible to a limited extent. In so-called **polar body diagnostics**, the genetic material of the polar bodies is examined. These are formed during meiosis (Sect. 4.1) and adhere to the egg cell for some time before they are broken down. Their genetic material is very similar to that of the egg cell—but not identical. Since sperm or egg cells are destroyed during a genetic analysis according to the state of the art, precise **pre-fertility diagnostics** is not possible at present. However, this is likely to change soon. Various research groups are reporting initial successes in the generation of germ cells from germ cell stem cells (Fig. 8.5). If these could be multiplied (cloned) in the laboratory, more precise diagnostics would be possible. What would be the conceivable effects?

In the case of genetic testing before fertilisation, there would be a selection of sperm and eggs if necessary. What are or would be the criteria? How do we separate a serious disease that would allow **selection** from a disease that causes discomfort at best? How do we deal with the probabilities of a genetic information actually taking effect during the life of the offspring? In the USA, it is already possible to select sperm cells by catalogue: for example, from Nobel Prize winners, from blue-eyed people or from successful sportsmen. What about **intelligence** or, rather, selection for cognitive traits? In 2017, a genetic study of 78,308 individuals identified 336 nucleotide positions (SNPs) distributed across 22 genes that are associated with cognitive traits [44]. Moreover, these SNPs are even related to Alzheimer's, depression, autism, and life expectancy, for example. As a rule, we want to send our children to the best school in place—so why not give our offspring the best genome possible? This may all sound very far away, but the technical possibilities already exist today. Ultimately, however, there are also opportunities after birth and at an advanced age to intervene in the genome in a therapeutic (Sect. 5.3) or optimising way in the course of gene therapy.

References

1. Sanger F, Coulson AR (1975). A rapid method for determining sequences in DNA by primed synthesis with DNA polymerase. J Mol Biol 94: 441–8. doi:https://doi.org/10.1016/0022-2836(75)90213-2

2. Maxam AM, Gilbert W (1977). A new method for sequencing DNA. Proc Natl Acad Sci USA 74: 560–4. doi:https://doi.org/10.1073/pnas.74.2.560

3. Ronaghi M, Karamohamed S, Pettersson B, Uhlén M, Nyrén, P (1996). Real-time DNA sequencing using detection of pyrophosphate release. Anal Biochem 242: 84–9. doi:https://doi.org/10.1006/abio.1996.0432

4. Shendure J, Ji H (2008) Next-generation DNA sequencing. Nat Biotechnol 26: 1135–1145. doi:https://doi.org/10.1038/nbt1486

5. Wheeler DA, Srinivasan M, Egholm M, et al (2008) The complete genome of an individual by massively parallel DNA sequencing. Nature 452: 872–876. doi:https://doi.org/10.1038/nature06884

6. Levy S, Sutton G, Ng PC, et al (2007) The diploid genome sequence of an individual human. PLoS Biol 5: e254. doi:https://doi.org/10.1371/journal.pbio.0050254

7. Venter JC (2008) A Life Decoded. Penguin Press, London/UK

8. Feng Y, Zhang Y, Ying C, et al (2015) Nanopore-based fourth-generation DNA sequencing technology. Genomics, Proteomics Bioinf 13: 4–16. doi:https://doi.org/10.1016/j.gpb.2015.01.009

9. Editorial (2018) The long view on sequencing. Nat Biotechnol 36: 287–287. doi:https://doi.org/10.1038/nbt.4125

10. Chinappi M, Cecconi F (2018) Protein sequencing via nanopore based devices: a nanofluidics perspective. J Phys: Condens Matter 30: 204002. doi:https://doi.org/10.1088/1361-648x/aababe

11. Karki R, Pandya D, Elston RC, Ferlini C (2015) Defining "mutation" and 'polymorphism' in the era of personal genomics. BMC Med Genomics 8: 37. doi:https://doi.org/10.1186/s12920-015-0115-z

12. Sudmant PH, Rausch T, Gardner EJ, et al (2015) An integrated map of structural variation in 2,504 human genomes. Nature 526: 75–81. doi:https://doi.org/10.1038/nature15394

13. Abel HJ, Larson DE, Chiang C, et al (2018) Mapping and characterization of structural variation in 17,795 deeply sequenced human genomes. bioRxiv 508515. doi:https://doi.org/10.1101/508515

14. Lin Z, Owen AB, Altman RB (2004) Genomic research and human subject privacy. Science 305: 183–183. doi:https://doi.org/10.1126/science.1095019

15. Weber-Lehmann J, Schilling E, Gradl G, et al (2014) Finding the needle in the haystack: Differentiating "identical" twins in paternity testing and forensics by ultra-deep next generation sequencing. Forensic Sci Int: Genet 9: 42–46. doi:https://doi.org/10.1016/j.fsigen.2013.10.015

16. Dobzhansky T (1973) Nothing in Biology Makes Sense except in the Light of Evolution. Am Biol Teach 35: 125–129. doi:https://doi.org/10.2307/4444260

17. Pääbo S (2014) The Human Condition—A Molecular Approach. Cell 157: 216–226. doi:https://doi.org/10.1016/j.cell.2013.12.036

18. Krause J, Trappe T (2019) Die Reise unserer Gene: Eine Geschichte über uns und unsere Vorfahren. Propyläen Verlag, Berlin

19. Underwood FM, Burn RW, Milliken T (2013) Dissecting the Illegal Ivory Trade: An Analysis of Ivory Seizures Data.

PLoS One 8: e76539. doi:https://doi.org/10.1371/journal.pone.0076539

20. Green RE, Krause J, Briggs AW, et al (2010) A Draft Sequence of the Neandertal Genome. Science 328: 710–722. doi:https://doi.org/10.1126/science.1188021

21. Reich D, Green RE, Kircher M, et al (2010) Genetic history of an archaic hominin group from Denisova Cave in Siberia. Nature 468: 1053–1060. doi:https://doi.org/10.1038/nature09710

22. Slon V, Mafessoni F, Vernot B, et al (2018) The genome of the offspring of a Neanderthal mother and a Denisovan father. Nature 561: 113–116. doi:https://doi.org/10.1038/s41586-018-0455-x

23. Fleckinger A (2018) Ötzi, the Iceman. Folio Verlag, Bozen/IT

24. Keller A, Graefen A, Ball M, et al (2012) New insights into the Tyrolean Iceman's origin and phenotype as inferred by whole-genome sequencing. Nat Commun 3: 698. doi:https://doi.org/10.1038/ncomms1701

25. Handt O, Richards M, Trommsdorff M, et al (1994) Molecular genetic analyses of the Tyrolean Ice Man. Science 264:1775–1778. doi:https://doi.org/10.1126/science.8209259

26. Schierding WS, Cutfield WS, O'Sullivan JM (2014) The missing story behind Genome Wide Association Studies: single nucleotide polymorphisms in gene deserts have a story to tell. Front Genet. doi:https://doi.org/10.3389/fgene.2014.00039

27. Kremer LS, Bader DM, Mertes C, et al (2017) Genetic diagnosis of Mendelian disorders via RNA sequencing. 8: 15824. doi:https://doi.org/10.1038/ncomms15824

28. Cummings BB, Marshall JL, Tukiainen T, et al (2017) Improving genetic diagnosis in Mendelian disease with transcriptome sequencing. Sci Transl Med 9: eaal5209. doi:https://doi.org/10.1126/scitranslmed.aal5209

29. Dar S, Lazer T, Shah PS, Librach CL (2014) Neonatal outcomes among singleton births after blastocyst versus cleavage stage embryo transfer: a systematic review and meta-analysis. Hum Reprod Update 20: 439–448. doi:https://doi.org/10.1093/humupd/dmu001

30. Greely HT (2016) The End of Sex and the Future of Human Reproduction. Harvard University Press, Cambridge, Massachusetts/USA

31. Wilkinson J, Bhattacharya S, Duffy J, et al (2018) Reproductive medicine: still more ART than science? BJOG 126: 138–141. doi:https://doi.org/10.1111/1471-0528.15409

32. Millais JE (1889) The Theory and Practice of Rational Breeding. The Fancier's Gazette 11: 97

33. Millais JE (1884) An "Artificial Impregnation" by a Dog Breeder. Vet J 18: 256

34. Worboys M, Strange JM, Pemberton N (2018) The Invention of the Modern Dog: Breed and Blood in Victorian Britain. Johns Hopkins University Press

35. Jacobson CB, Barter RH (1967) Intrauterine diagnosis and management of genetic defects. Am J Obstet Gynecol 99: 796–807. doi:https://doi.org/10.1016/0002-9378(67)90395-x

36. Ceballos FC, Álvarez G (2013) Royal dynasties as human inbreeding laboratories: the Habsburgs. Heredity 111: 114–121. doi:https://doi.org/10.1038/hdy.2013.25

37. Ceballos FC, Joshi PK, Clark DW, et al (2018) Runs of homozygosity: windows into population history and trait architecture. Nat Rev Genet 19: 220–234. doi:https://doi.org/10.1038/nrg.2017.109

38. Helgason A, Palsson S, Guthbjartsson DF, et al (2008) An Association Between the Kinship and Fertility of Human Couples. Science 319: 813–816. doi:https://doi.org/10.1126/science.1150232

39. Edmands S (2006) Between a rock and a hard place: evaluating the relative risks of inbreeding and outbreeding for conservation and management. Mol Ecol 16: 463–475. doi:https://doi.org/10.1111/j.1365-294x.2006.03148.x

40. Ebenesersdóttir SS, Sandoval-Velasco M, Gunnarsdóttir ED, et al (2018) Ancient genomes from Iceland reveal the making of a human population. Science 360: 1028–1032. doi:https://doi.org/10.1126/science.aar2625

41. Wojcicki A, Avey L, Mountain JL, et al (2013) Free Patents Online. freepatentsonline.com/8543339.pdf

42. Cao A, Kan YW (2013) The Prevention of Thalassemia. Cold Spring Harbor Perspect Med 3: a011775–a011775. doi:https://doi.org/10.1101/cshperspect.a011775

43. Carmi S, Hui KY, Kochav E, et al (2014) Sequencing an Ashkenazi reference panel supports population-targeted personal genomics and illuminates Jewish and European origins. Nature Comm 5: 119. doi:https://doi.org/10.1038/ncomms5835

44. Sniekers S, Stringer S, Watanabe K, et al (2017) Genome-wide association meta-analysis of 78,308 individuals identifies new loci and genes influencing human intelligence. Nat Genet 49: 1107–1112. doi:https://doi.org/10.1038/ng.3869

5
Editing Genetic Material

In the preceding sections we have become acquainted with
cases in which genetic diagnostics led to a reduction of hereditary diseases in population groups. In addition to the
public health benefit, the control of reproduction is also
justified by some on the grounds of financial relief for the
community. Without evaluating this at this point, it must
be clear that if this leads to abortion or, *in vitro,* the disposal
of an embryo, life is sacrificed. Morally exculpatory, in some
cases, it can be argued that the interventions serve to protect the health of the pregnant woman. Today we are faced
with the situation that we can edit genes, that is correct
them. This in itself is nothing new. In Chap. 3 we have already become acquainted with various methods for intervening in the genetic material of living beings. In general, a
distinction must be made between non-directed methods,
which induce genetic changes at random either via ionising
radiation (Sect. 3.1) or via chemicals, and specific or site-
directed methods. The latter procedures are gaining increasing interest and are now the talk of the scientific community as **genome editing** or gene surgery. As a result of the
increasing clarification of the relationship between the

© Springer-Verlag GmbH Germany, part of Springer Nature 2022
R. Wünschiers, *Genes, Genomes and Society,*
https://doi.org/10.1007/978-3-662-64081-4_5

appearance (**phenotype**) of an organism and its genetic make-up (**genotype**), one is no longer dependent on chance. On the contrary, one would like to intervene specifically in the genome. An important milestone in modern molecular biology was the first description of *site-directed mutagenesis* using short specific DNA sequences (the so-called *primer-extension method*) in 1978. The English biochemist Michael Smith was awarded the Nobel Prize in 1993 for establishing this technique. Various subsequent methods became more and more sophisticated and precise. But the last major breakthrough came with the application of the CRISPR/Cas system, as first described in 2012. It is currently fuelling a global bioethical debate [1–3]. By means of genetic intervention, we can therefore extend the argument of health integrity to the embryo and, if a genetic disease is diagnosed, cure it through gene therapy by interfering with its genetic material.

5.1   CRISPR/Cas System

In August 2012, the US biochemist Jennifer Doudna from the *California University* in Berkeley and the French microbiologist Emmanuelle Charpentier, now director at the *Max Planck Institute for Infection Research in* Berlin, published a ground-breaking work—economists would probably say *game changer*: a method for the targeted modification of a specific DNA sequence in the genome [4]. A few months later, in February 2013, the US biochemist Feng Zhang published how the system can be applied to human and mouse cells [5]. The method is based on the CRISPR/Cas system and can induce nucleotide-precise changes in the genome, as precisely as a scalpel on an surgical table. This is why the procedure is also known as gene surgery or, in reference to the genetic code, **gene editing**. Today it is clear that gene editing work on any living organism. The procedure is so simple that it is even sold to everyone in the USA as an experimental kit for currently US$159 and can be used in schools (Sect. 7.1). And last but not least: The CRISPR/Cas system has sparked a **dispute** between representatives from science and industry and regulatory authorities as to whether or not living organisms treated with it should be assessed as genetically modified organisms and thus regulated. The **European Court of Justice** (ECJ) ruled in its judgment of July 25, 2018, that organisms developed with the help of gene editing must be regulated like all other genetically modified organisms. I have explained the reasons why this ruling is now the subject of heated debate in Sect. 3.4. Of course, this discussion is important, but was caught up with reality on November 27, 2018. On that day, Chinese biophysicist He Jiankui announced that he had gene edited human embryos, that is, genetically modified them, and that they had already been born at the time of the announcement. Although many scientists had suspected, feared, hoped, predicted it, the announcement hit like a bomb.

But first things first. **CRISPR** stands for *clustered regularly interspaced short palindromic repeats.* Whew, what is that? They are contiguous *(clustered)* palindromic sequences that are repeated and *interspaced* at equal intervals *(regular).* A simple palindrome is racecar: it can be read from the front and the back. In genetics, a palindromic sequence indicates that it has the same sequence of characters on the opposite strand of DNA in the reverse direction. For example, the sequence AACGTT reads the same on the opposite strand. What does it mean that such DNA sequences occur in a genome? They were discovered in bacteria as early as 1987, when Japanese molecular biologist Ishino

Yoshizumi of Osaka University wrote in a paper, *an unusual [DNA] structure was found* [6]. It was a passing remark. The Spanish microbiologist Francisco Mojica described this unusual DNA sequence in more detail in 1993 and found out about 10 years later that the repeating palindromic sequences were only the separators for DNA sequences that bear great resemblance to the genetic material of known bacterial viruses [7–9]. In 2002, Dutch microbiologist Ruud Jansen, along with Mojica, proposed the name CRISPR [10]. While repetitive sequences were of no interest to most scientists, research was now picking up steam. In 2006, the US bioinformatician Eugene Koonin published a comprehensive study on CRISPR, showing for the first time how widespread they are in bacteria [11]. He developed the hypothesis that it could be a bacterial immune system against bacterial viruses (phages) (Fig. 5.1) [12].

Just as our immune system remembers which substances (antigens) it has already come into contact with, sequences

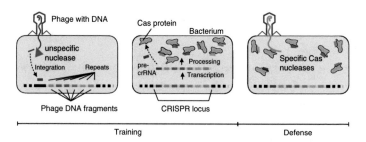

Fig. 5.1 (Left) A bacterium is attacked by an unknown phage. It injects its DNA into the cell. The phage DNA is cut non-specifically by nucleases. Resulting DNA fragments can be inserted into the CRISPR locus in the bacterial genome. (Middle) The CRISPR locus is read like a gene. The transcript is unravelled and the resulting CRISPR RNA (crRNA) is bound by Cas proteins. (Right) When the bacterium is infected by a phage known to it, the Cas nucleases (gene editing) can cut the phage DNA. Specificity is achieved by the bound crRNA molecules

of phages with which the bacterium, or a predecessor, has already had contact are stored in the CRISPR region. Food scientists first successfully confirmed this hypothesis with yogurt bacteria in 2007 [13]. Subsequently, several scientific teams reported that CRISPR-associated genes (abbreviated **Cas**) encode proteins that can cut DNA sequences. These Cas nucleases are the actual molecular **gene editors**. It is primarily thanks to the research collaboration between the groups around Doudna and Charpentier that the complete mechanism has been elucidated and optimised to such an extent that it can be adapted to all living organisms. Today, we know that there are three different CRISPR immune systems in bacteria and that there are many variations of these [14]. The system that has been developed to biotechnological maturity is based on type II, which I will therefore restrict myself to here.

According to this, a short fragment of the phage's double-stranded DNA (called a spacer), usually 20 base pairs long, is inserted into a special area on the chromosome of the infected bacterium. This area is called the CRISPR locus. The phage DNA fragment is delineated from pre-existing fragments by the CRISPR sequence (repeat) (Fig. 5.1). The information about the phage infestation is thus stored. In fact, scientists have already modified the CRISPR system so that certain events, such as the detection of a chemical, are stored in the genome [15]: With each cell division, the information is passed on to the offspring. In addition, one CRISPR RNA molecule (crRNA) is formed from each of the phage DNA fragments in the CRISPR locus. This first combines with a tracrRNA to form a cr/tracrRNA hybrid called a **guideRNA**. This then forms a joint complex with a Cas nuclease (Fig. 5.2).

While the tracrRNA and the Cas nuclease are always identical and thus universal, the crRNA, derived from the

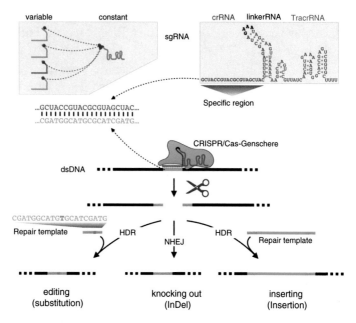

Fig. 5.2 The CRISPR/Cas system is based on a synthetic guide RNA molecule (sgRNA, single guide RNA) consisting of a variable crRNA (CRISPR RNA) and a tracrRNA (trans activating crRNA), which are connected to each other via a bridge (linkerRNA). The crRNA works like a probe and directs the Cas nuclease to the target sequence on the double-stranded DNA (dsDNA). After the DNA is cut, it is reassembled by various repair mechanisms (HDR: homology directed repair; NHEJ: non-homologous end joining). This repair can be influenced by added DNA molecules (repair templates). NHEJ leads to an inaccurate change in the gene sequence (with an insertion or deletion) and thus usually to silencing of the gene. The HDR mechanism can be used to selectively replace base pairs (substitution) or insert whole genes (insertion). Note that unlike DNA, RNA has a T (thymine) replaced by a U (uracil). Like the T, the U also pairs with A (adenine)

phage genome, is specific. The energetic price that a bacterium has to pay for specific phage defence is the constant biosynthesis of the three components of the defence system. If a bacterial cell is attacked by a phage from which a DNA fragment is already deposited in the CRISPR locus, then the Cas nuclease appropriately loaded with a cr/tracrRNA hybrid can specifically bind to the phage DNA and cleave it at the binding site (Fig. 5.2). For biotechnological applications, it was an important discovery that crRNA and tracrRNA can be linked via a short linkerRNA, that is a linking RNA molecule. This synthetic construct is called sgRNA (single guide RNA). This can be produced cheaply chemically and is usually ordered by scientists from service providers. All that is required is the sequence of the specific region. This depends on the gene sequence that is to be cut.

Cutting DNA? That does not sound good at first. In fact, repair mechanisms immediately become active in a cell when the hereditary molecule is cut. These **repair mechanisms** are universally present in every known living thing in every cell, because DNA strand breaks are nothing unusual. They can be caused by UV light or ionizing radiation, for example. The fact that every cell can reassemble cut DNA means that the CRISPR/Cas system is also universally applicable. Two fundamentally different repair mechanisms can be distinguished, homologous recombination (HDR, homology directed repair, Fig. 5.2) and non-homologous repair (NHEJ, non-homologous end joining, Fig. 5.2).

In **homologous recombination,** a second chromosome serves as a template, usually the intact daughter chromosome (Sect. 2.1). By injecting a high concentration of DNA fragments, the probability increases that these will be used as a template for repair instead of the sister allele. It is only important that the ends of the added DNA fragments match (are homologous) to the ends of the cleavage site. They must be able to form base pairings. In this way, single base pairs can be edited or entire genes can be inserted (Fig. 5.2). In contrast, in **non-homologous repair**, the ends are fused without a template. In this process, errors occur, that is, additional or missing base pairings occur. We then speak of **InDels** (insertions or deletions). If a gene coding for a protein has been cut, the InDel is most likely to cause a shift in the reading frame of the genetic code (Fig. 2.4). Consequently, the resulting protein will no longer function correctly. The gene is said to have been *silenced* (gene silencing or knockout). These are the basic ways of using gene editing. However, there are still many variations. Just as there are different versions of scissors we use in everyday life (for example, spring scissors or pivoted scissors) for different types of material to be cut (such as paper, fingernails, fabrics or bones), there are also different nucleases for different DNA and RNA targets. Only recently, it was even shown that thousands of positions can be edited at once using a modified pair of gene editing (Sect. 6.2).

With the CRISPR/Cas system, the bacterial immune system has become the most effective genetic engineering tool currently available [1]. The CRISPR/Cas system is more effective, cheaper and easier to adapt than the **ZFN** (zinc-finger nucleases) available since 1996 and the TALEN (transcription activator-like effector nucleases) applied since 2010. It is quite remarkable that after the revolution of life sciences by restriction enzymes (Sect. 3.4) in the 1970s, again a bacterial defence mechanism against phages makes furore in the 2010s.

Are there also **problems**? Yes, it can happen that the CRISPR/Cas system makes mistakes and additionally induces changes at other locations in the genome. We refer to this as **off-target effects**. What this is related to is not always entirely clear. The probability that the same nucleotide sequence of 20 letters will occur a second time in a genome and that the gene editing will therefore bind and cut there is very low: for a twenty-letter word of four characters (A, T, G and C), it is 1:1,099,511,627,776, or about one in a trillion. Minimizing off-target effects is currently an important goal of worldwide research. Another important goal is to reliably detect off-target effects. Current developments in DNA sequencing in general and single-cell DNA sequencing in particular are helping in this regard (Chap. 4).

Curiously, another problem is the immunity of organisms to the bacterial immune system, or more precisely, to the Cas protein [16]. Various research teams have discovered **antibodies** against the Cas nuclease in the human immune system [17]. In the case of gene therapy, this could not only inactivate the nuclease, but could also lead to an immune shock. Here, too, research is being carried out to find solutions. In addition, it has not yet been definitively clarified how other cell components react to the CRISPR/Cas system. However, these problems only arise at the moment of application. Once the DNA has been edited, the system is lost in subsequent cell divisions.

A particular risk when using gene editing on fertilized egg cells or embryos is **mosaicism**. In most applications, three components (the Cas nuclease, the sgRNA and a repair template) are injected into the cells (Fig. 5.2). In 2015, a team led by Chinese stem cell researcher Junjiu Huang was the first to use gene editing on human embryos [18]. They did so by using so-called **tripronuclear (3PN) zygotes**, which cannot develop into complete embryos. They are created during *in-vitro fertilization* when two sperm cells instead of one enter the egg cell. This occurs in about 4% of all cases. Therefore, they are usually destroyed. But precisely because of their inability to develop into embryos, they lend themselves to embryo research on ethical grounds. The researchers injected the zygotes, which already consisted of several cells, with sgRNA, a Cas9 mRNA (which provides the Cas nuclease after translation), and a single-stranded DNA that serves as a proofreading template (see previously). The sgRNA targeted the HBB gene responsible for β-thalassemia (Fig. 2.3 and Sect. 4.3). Out of a total of

86 treated embryos, 71 embryos survived, of which again 59 contained all injected components. Among these, in 28 embryos the Cas nuclease cut correctly. In seven embryos, a chromosome was used as a repair template rather than the injected DNA. In four embryos, the HBB gene was correctly edited with the DNA template. However, these four embryos were a mosaic of cells that were repaired in different ways. This means that not all cells were edited after injection of the gene shearing components. So, the embryos consisted of both edited and non-edited cells. This in itself is not a bad thing. In a sense, we are all genomic mosaics, since isolated mutations can occur in stem cells, for example, which all descendant cells then carry as well. Women are also mosaics, in whose cells it is randomly determined after each cell division which of the two X chromosomes will be inactivated. If both were active, severe damage would occur. However, the appearance of mosaics after treatment with the CRISPR/Cas system shows that the attempt to repair a genetic disease can also end incompletely.

The most controversial experiment on human embryos was conducted by Chinese biophysicist He Jiankui: He gene edited **human embryos** with the CRISPR/Cas system, which were then born as the twins Nana and Lulu in 2018. I discuss the social implications of this in Sect. 5.2. Here, I will first discuss the experiment itself. So far, hardly anything has been published about it. The primary basis is therefore He's presentation entitled *CCR5 gene editing in mouse, monkey and human embryos using CRISPR/Cas9*, which he gave on November 28, 2018 at the *Second International Summit on Human Genome Editing in* Hong Kong—and much that has been written about it since then [19]. There is no scientific publication on this by He himself, and it is a big question whether there ever will be (Sect. 5.2).

What gene editing did He do? I have to backtrack a bit on this. For AIDS patients, 2009 was a year of great hope: the *Charité Hospital* in Berlin announced that the so-called

Berlin patient had been cured of AIDS following a bone marrow transplant [20]. The Berlin patient was Timothy Ray Brown, a 40-year-old American at the time. After his AIDS infection had been stably suppressed with medication for several years, he was diagnosed with a serious blood cancer (acute myeloid leukaemia) in 2006. As a therapy, the physician Dr. Gero Hütter of the *Charité Hospital* enabled him in February 2007 to receive hematopoietic stem cells from a donor who is a homozygous carrier of the **CCR5Δ32 variant** of a receptor protein and thus immune to the HI virus [21]. The therapy worked and was successfully repeated 10 years later in another patient, the **London patient** [22]. The *CCR5* gene encodes a chemokine receptor that serves as an entry site for the most widespread and aggressive HI virus (type R5-tropic HIV-1), together with the CD4 glycoprotein (Sect. 7.1) on the white blood cells (T cells) of the immune system (Fig. 5.3).

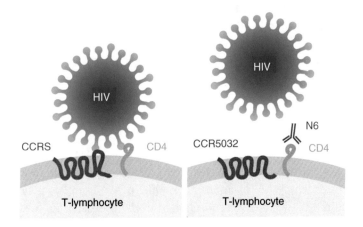

Fig. 5.3 (Left) The HI virus infects a white blood cell of the immune defence. To do this, it must bind to the CCR5 receptor and the CD4 glycoprotein. (Right) The absence of 32 base pairs (Δ32) in the *CCR5* gene causes the virus to no longer recognize the protein. The N6 antibody can be used to mask the CD4 glycoprotein. Both mechanisms lead to immunity against HIV

In the *CCR5Δ32* variant, 32 base pairs are missing—the delta sign stands for a deletion. This leads to resistance to the HI virus if the gene is homozygous, that is both the maternal and the paternal allele contain this deletion. The allele probably originated about 2000 years ago at the time of the Vikings. In Europe, the proportion of homozygous carriers is 1–2%. Interestingly, the HI virus in its human-infecting form is still relatively young. It is believed that the first transmissions from chimpanzees to humans occurred around 1900 in Cameroon [23]. In contrast, the *CCR5Δ32* variant of the receptor protein first appeared in humans about 2000 years ago, presumably in the region of what is now Finland, and has apparently been positively selected for since then [24]. This means that it gives carriers of this allele an advantage and has therefore spread since that time. Scientists suspect that the advantage has been resistance to smallpox viruses [25].

He's motivation for his gene therapy in the human germ line (Fig. 8.5) is primarily that couples with AIDS can have offspring. The risk for new-borns to contract AIDS is increased many times in the first months of life. By inserting a deletion in the *CCR5* gene in fertilized eggs, He hopes to make the babies immune to infection with HI viruses. He recruited seven couples from an AIDS support group for the gene therapy, in which the man has AIDS but the woman does not. The couples and babies were reimbursed for treatment and follow-up costs. This corresponds to an equivalent value of about €35,000. In two couples, gene-edited blastocysts were implanted in the mother (Fig. 5.4).

The main focus of the research team led by He Jiankui was to avoid off-target editing as described above. Therefore, the parents as well as the blastocysts and the babies were completely sequenced several times. For genome sequencing (Chap. 4), He used single-cell sequencing, in which a

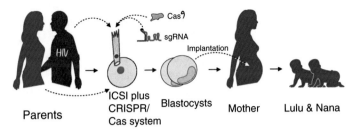

Fig. 5.4 The twins Lulu and Nana are the first gene edited babies. During *in-vitro* fertilization or, more precisely, intracytoplasmic sperm injection (ICSI), they were injected not only with the sperm but also with the protein of the DNA nuclease Cas9 and a *single guide* RNA (sgRNA). Cas9 and the sgRNA together form the gene editing (Fig. 4.11)

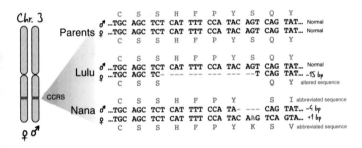

Fig. 5.5 Sections of the DNA sequences of the *CCR5* gene on chromosome 13 in the parents and their twins Lulu and Nana. Only one DNA strand of each chromosome pair is shown. Above and below the DNA sequences, respectively, the amino acid sequences (as single letter codes) of the resulting CCR5 protein are shown. The deletions and insertions in the DNA are marked in green

single cell is sufficient to read the entire genome. Only about five cells can be taken from the blastocyst before implantation without affecting its development. No off-target edits were detected in the babies. However, the alterations of the *CCR5* gene are different in the two twins (Fig. 5.5).

While Lulu's genome contains an unchanged version (allele) and a version with a 15 base pair deletion, in Nana both alleles are changed: One allele contains a four base pair deletion while the other allele contains an insertion of one base pair. From these results it follows that Lulu, with only one altered allele, is heterozygous and can be infected by HIV. The intact *CCR5* gene has a dominant effect (Fig. 4.6). Thus, gene therapy was unsuccessful here. Nana, on the other hand, is expected to be resistant to AIDS. This raises the question of how the twins and also the parents will cope with the fact that only one child carries the desired trait? In addition, the procedure shows that at least three different alleles have been created. This is due to the procedure in which no correction template was used (Fig. 5.2), but only the imprecise DNA repair system. So, the Cas nuclease cut precisely and the edits are all within the range targeted by Hes team. But, as expected, the repair was not exact. The twins are due to have regular medical check-ups until they are eighteen.

In addition to Lulu and Nana, another edited baby is expected to be born to another couple in the summer of 2019. All other embryos created will not be used for now, as Chinese authorities have banned He from continuing. It is important to note that He's experiment violated Chinese law, which specifically prohibits the implantation of embryos generated in research (Sect. 5.2). And in June 2019, Russian molecular biologist Denis Rebrikov announced that he would also create HIV-resistant CRISPR babies at his reproduction clinic in Moscow.

A major criticism of He's gene therapy was also its goal. AIDS is a serious disease, but it can be prevented in other ways. Most scientists thus regard the intervention more as an **enhancement** (Sect. 7.3) than as a therapy. However, the risks taken by He are disproportionately high. In addition, **other functions** are attributed to the *CCR5* gene that may be impaired by the deletion. For example, carriers of defective *CCR5* alleles appear to be more susceptible to infection by other viruses such as West Nile fever virus, which can cause a form of meningitis. There is also evidence that the CCR5Δ32 variant of the chemokine receptor protein contributes to a more severe, possibly fatal, course of common flu infection [26]. In addition, the *CCR5* gene appears to play a role in the development of cognitive processes [27]. For example, it has been shown in mice that impairment of molecular signal transduction via the CCR5 receptor protein leads to improved memory conduction [28]. However, what the effects are in humans is still unexplored. Thus, in addition to ethics, there are also many medical reasons for keeping our hands off the *CCR5* gene in the germ line at the present time (Sect. 7.1).

Is gene editing now genetic engineering? This question can be answered with a resounding yes, if one looks at the procedure. The tool for genetic mutagenesis must be introduced into the target cell, that is a Cas enzyme (usually Cas9) as a DNA-cutting nuclease (or alternatively the coding mRNA) and the sgRNA construct that directs the enzyme to the correct position on the DNA. There are several ways to do this. In the simplest case, the Cas protein and the sgRNA are introduced directly into a cell. They do their work and are rapidly degraded in the cell. In most cases, however, the corresponding genetic sequence information is integrated into a DNA vector (plasmid, virus), which is then introduced into the target cell. In both processes, genetic information is integrated into a cell. In plants and bacteria this is called **transformation**, in vertebrates such as humans it is called **transfection**; and it is a genetic engineering process. However, the vector always remains separated and is not integrated into the genome. It is merely expressed, whereupon the CRISPR/Cas system does its work. The DNA vector is then lost in the daughter cells of the subsequent cell division. It is not duplicated and passed on like the genetic material, but is only present temporarily (transiently). It can therefore be said that the process is genetic engineering, but the product is not. In its ruling, however, the **ECJ focused on the process**. Irrespective of this important regulatory issue, gene editing is currently being used intensively in medical and breeding research, among other areas. There is a gold-rush atmosphere.

5.2   China's CRISPR Crisis?

While we not only praise 2500-year-old traditional Chinese medicine at esoteric fairs, but also award its methods in the form of malaria treatment with artemisinic acid from the mugwort plant (Sect. 6.2) with the Nobel Prize (2015), research into the future of Chinese medicine is being carried out in China's biomedical laboratories. For example, also in 2015, the first experiments with CRISPR/Cas system were conducted on human embryos. And in 2018, Chinese biophysicist He Jiankui of Shenzhen's *Southern University of Science and Technology* made the oft-thought-of, poetized, and filmed reality—putting China at the centre of an international debate on how to deal with genetic engineering in general and human germline gene editing in particular. Several cautionary critics of the use of gene therapy in the germline had suspected that the first gene-edited babies would be born in a private fertility clinic in, possibly Mexico (Sect. 5.3, Three-Child Parents). And indeed, it is not known whether there are more or less illegal cases of the use of gene therapy at such clinics (Fig. 5.6). Public is the case of He from China and it is worth reflecting on whether this is coincidence or systemic.

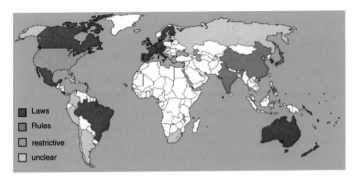

Fig. 5.6 Existing laws and regulations relating to the use of the CRISPR/Cas system in the human germline. States marked in white were not studied. (Sources: Ishii (2017), Ledford H (2015) [29, 30])

He is responsible for the birth of the first genetically modified twins in autumn 2018 (Sect. 5.1). It remains to be seen how the parents and their children will deal with this burden, how harassed they will become. They still live in anonymity and the events still seem almost like a dream. Or nightmare? In fact, as of this writing, there is still no scientific evidence of the work He describes. Is it all just fake news? Although one would hope so, no, the red line has indeed been crossed. This is clear, among other things, from the Chinese government's handling of the spectacular case. Even if China is said to be able to do almost anything with genetic engineering, there are rules (Fig. 5.6).

Transplanting embryos generated in research into the uterus and carrying them to term has been expressly prohibited in China since 2003. Thus, He has not only violated the warning repeatedly issued by various international scientific organizations and government agencies not to perform genetic engineering applications on human embryos. He has also violated **Chinese laws** and will be prosecuted accordingly. Remarkably, He prefaced his presentation of the results on Lulu and Nana at the *Second*

International Summit on Human Genome Editing in Hong Kong in November 2018 with an apology. He shared that his university had no knowledge of the work presented. In parallel with his experiments, He published a draft of five ethical guidelines for gene therapy in support of reproductive technologies such as *in-vitro* fertilization [31]:

- **Mercy for families in need**
  A defective gene, infertility or preventable disease should not extinguish life or undermine the union of a loving couple. For some families, early gene therapy may be the only viable way to cure a hereditary disease and save a child from suffering their life.
- **Only for serious disease, never vanity**
  Gene therapy is a serious medical procedure that should never be used for purposes of aesthetics, enhancement, or sex selection-or in any way that would interfere with a child's well-being, enjoyment, or free will. No one has the right to determine a child's genetics except to prevent disease. Gene therapy exposes a child to potential safety risks that may be permanent. Gene therapy may be performed only if the risks of medical intervention outweigh the risks of gene therapy.
- **Respect a child's autonomy**
  Life is more than our physical body and its DNA. After gene therapy, a child has the same right to a free life, his profession, his citizenship and his privacy. There are no obligations, even financial, to his parents or to any organization.
- **Genes do not define you**
  Our DNA does not define our purpose or what we might achieve. We flourish through our hard work, nourishment and the support of society and our loved ones. Whatever our genes, we are equal in dignity and potential.

- **Everyone deserves freedom from genetic disease**
  Wealth should not determine health. Organizations that develop genetic cures have a deep moral obligation to serve families of all backgrounds.

This publication was withdrawn by the journal after his work became known. However, they give an insight into He's thinking, although one must naturally take his **socialization** into account. There is an excellent article on this by Chinese particle physicist Yangyang Cheng from the USA [32]. He's bioethics guidelines are quite general. Apart from the fact that he is not a bioethicist but a physicist, the nature of the Chinese language certainly plays a role here [33]. It is ambiguous and does not fit in at all with the clear structure of scientific linguistic culture. The translation of the language, which is difficult to master anyway, into English can therefore only limp along and must be taken with a grain of salt.

In response to He's announcements, the US philosopher and bioethicist Sheldon Krimsky of *Tufts University* lists ten bioethical principles that He violated with his experiment [34]. Among his criticisms is that He is a physicist, not a medical doctor, and thus lacks a reputation and experience in handling and evaluating experiments on humans. This is an aspect that we encounter again in the topic of citizen science (Sect. 7.1). Also, He did not inform his university about his experiments. Krimsky also accuses He of acting in his own interest, as he has stakes in several biotech companies, which calls his neutrality into question. Similarly, he accuses He of recruiting couples out of personal and emotional distress and with a high financial consideration (around €35,000). In addition, he claims that he inadequately informed his subjects about the risks of the experiments. At the heart of this issue is the consent form, which Krimsky says is too complex and technically worded. It also does not provide information about alternative treatments for AIDS.

The auxiliary bishop of Augsburg, Anton Losinger, who was a member of the German Ethics Council from 2008 to 2016, accuses He of violating fundamental human rights. Losinger argues that, firstly, human dignity is being attacked at its deepest core because the babies' future descendants are also affected. Secondly, he questions the aim of gene therapy and establishes a link with eugenics. Finally, like Krimsky, he accuses He of conducting research in the field of human gene therapy primarily for economic reasons.

Ethicist Bettina Schöne-Seifert warns of the massive impact that He's approach could have on the reputation of genetic engineering in clinical use [35]. She points to the need to draw a line between therapy and the spectre of enhancement, but also to the absurdity of many categorical arguments for banning it. These often refer to the idea that the human gene pool must remain intact. Yet even the global **UNESCO Declaration** *On the Protection of the Human Genome* explicitly describes in Article 3:

> The human genome, which by its nature evolves, is subject to mutations.

and in Article 5a:

> Research, treatment or diagnosis affecting an individual's genome shall be undertaken only after rigorous and prior assessment of the potential risks and benefits pertaining thereto and in accordance with any other requirement of national law.

Indeed, the germline application of gene editing by He's research team has damaged both China's reputation and gene editing as a method—but it has also helped through the dialogue that has ensued. I would like to briefly explore here the role played by China's political system and cultural background: Why were the first germline-edited babies born here, of all places?

The Chinese government hopes that the biotechnology revolution will do for it what Sputnik did for the Soviet Union and the moon landing did for the United States: Evidence of the strength of a political system in helping science achieve lofty goals. And from a European perspective, it must be said that China is a system rival [36]. The effort being expended is enormous and the willingness of the Chinese people to achieve the goals is, more or less forced, high. An insight is provided by the *Beijing Genome Institute* (BGI), founded in 1999 and based in Shenzhen in Guangdong Province, with field offices in Cambridge near Boston in the USA and Frederiksberg in Denmark. As an independent research institute, it initially participated in the international human genome project (Sect. 7.2) and over the years developed into one of the most influential genomics institutes in the world. In Germany, the BGI became known in 2011 through the rapid decoding of the genome of the pathogenic EHEC bacterium. This helped to develop a diagnostic test very quickly and thus contain the epidemic of the pathogen. The pathogen caused around 4300 illnesses and 50 deaths in the 2011 outbreak in Germany.

China's society and culture have been shaped by the peaceful teachings of Kong Qiu (Confucianism), Siddharta Gautama (Buddhism) and Laozi (Daoism) for over 2000 years. According to this philosophy, the individual must always serve the good of society and strive for education. The goal of the individual and society is to eliminate suffering from the world. From these schools of thought, together with the ruling political system, in which the Chinese Communist Party has authoritatively ruled since 1949, a society oriented towards **performance and**

obedience has been formed. From 1980 to 2015, the **one-child policy** was in place to prevent famine and enable economic progress. This resulted in a society of loneliness and parents we would call helicopter parents by Western standards. This has also resulted in the value of an embryo being valued less than in Western cultural circles, or in other words, if I have to limit myself in the number of offspring, then please let it be the best possible child. This provides the breeding ground on which the ideas of a healthy society in the sense of He flourish. Sponsored by the communist government, scientists try to contribute to the **welfare** of the state and social system to the brink of exhaustion. Reproduction clinics provide the raw material for embryonic development research and intervention. We are far from knowing everything that is being done in China, because the recognition of outstanding scientific work does not depend on **publications** in international journals. On the contrary, publishing only distracts from the work and the international reputation was usually obtained beforehand anyway during scientific training in the USA or Europe. Access to research funds is rather the other way round than we know it in the West. In our country, the state announces funding programmes in the context of which scientists apply for **research funding** with their ideas. In China, government-affiliated institutes select who or which institute will receive funding. The whole thing is mixed with a closely interwoven market economy, which then also accesses the international market. It so happens, for example, that in January 2019 it became known that a team of researchers from the *Chinese Academy of Sciences* and the *Research Center for Brain Science and Brain-inspired Technology,* both in Shanghai, reported in a Chinese magazine that they had edited macaque monkeys in the germ line with the CRISPR/Cas system and then cloned them, that is, duplicated them like Dolly the sheep (Sect. 3.5) [37, 38]. This was published in the journal *National Science Review,* which is published under the supervision of the *Chinese Academy of Sciences.*

He's work on Lulu and Nana has not yet been published. The only account of his approach available to the public is the talk he gave at the *International Summit on Human Genome Editing* on November 28, 2018. Nevertheless, we may assume that the representations are genuine. This is supported by the very fact that the authorities publicly distanced themselves from He's experiments, initiated an investigation and blocked all relevant postings on the very restrictive Chinese internet. This also applies to Internet pages in which He was praised for his achievements in the further development of single-molecule sequencing and which resulted in the founding of the company *Direct Genomics.* Scientists are also apparently encouraged not to talk about the process. China has suffered damage to its image and is trying to ride it out. Science journalist David Cyranoski, Asia correspondent for the scientific journal *Nature* in Shanghai, reports that users of the messaging service *WeChat* (a cross between *Facebook* and *WhatsApp*) have been prevented from sharing information about He's experiments.

It remains to be said, then, that the mix of the Chinese scientific system, society, and political directives created a framework in which He felt comfortable with his experiments and was able to defy internationally expressed concerns.

So much for the patterns of thought and action in China. In the West, on the other hand, the **Christian view** of interfering with the genetic make-up and the germ line shapes the way we act. And what conclusions are reached by **Muslims**, who, with around 1.6 billion believers, are the second largest religious community on the planet? In traditional Islamic teachings, there are five principles for addressing ethical issues [39]:

- **Principle of intention** *(Qasd):* It is fulfilled when it comes to sparing the twins suffering.
- **Principle of certainty** *(Yaqin):* This principle is questioned because it cannot be considered certain what the actual outcome of the treatment will be.
- **Principle of injury** *(Darar):* The cure and prevention of disease is obligatory; however, the balance between the chances and risks of treatment can hardly be weighed in the case of the twins, see *Yagin.*
- **Principle of necessity** *(Darura):* Necessity permits what is forbidden; in the case of the twins, however, the intervention was not necessary, since AIDS can be prevented and treated in other ways.
- **Principle of custom** *(Urf):* It is immoral to use a method about which a majority expresses misgiving.

It should be noted that since the Middle Ages, Islamic philosophy has provided an ethical compass in medicine upon which Western science has been built and is partially based today.

In March 2019, scientists from seven nations, including the discoverers of the CRISPR/Cas system Emmanuelle Charpentier and the further developer Feng Zhang, called for a **moratorium** [40]. In their article in the scientific journal *Nature,* they call for the immediate cessation of all research in which the gene editing is used to interfere with the germ line. They also call for a global monitoring body to ensure that minimum standards are obeyed. Work is to be suspended until this supervisory body, which countries are to join voluntarily, is in place. Although this moratorium once again shows that the scientific community as a whole is capable of self-regulation, it is questionable whether it will prevent individuals from using the technique. The former chairman of the German Ethics Council, Peter Dabrock, is calling for a worldwide supervisory authority for gene experiments on humans along the lines of the *International Atomic Energy Agency* (IAEA). In January 2019, the *World Health Organization* (WHO) established an *Expert Advisory Committee* on *Developing Global Standards for Governance and Oversight of Human Gene Editing*. The first inaugural meeting of the 18 members from 14 nations took place in Geneva on March 18, 2019. It remains to be seen what influence the new committees will have on the future of germline gene editing in humans.

5.3   Gene Therapy

The intervention in the genome in order to treat genetic diseases is called gene therapy. In September 1990, the first officially approved gene therapy was performed in the USA. The team led by the US physician French Anderson treated the four-year-old girl Ashanti Dasilva, who suffered from the hereditary, *severe combined immunodeficiency* (SCID). It is a monogenic disease, which means that a single defective gene is responsible for it. The doctors took blood from the child and, with the help of a virus, introduced a healthy copy of the gene (cargo gene) into white blood cells (T cells) (Fig. 5.7) [41]. The blood treated in this way was returned to the child by infusion. The therapy was successful, but not sustainable, as the transgenic cells did not settle stably in the tissue. Therefore, Ashanti had to be treated again at regular intervals.

In December 1992, the team led by oncologist Roland Mertelsmann from *Freiburg University Hospital* received

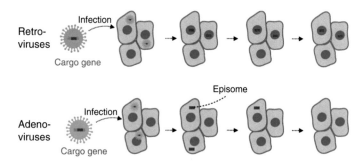

Fig. 5.7 Retroviruses and adenoviruses can be equipped with additional genetic information (cargo gene). When infecting a host cell, retroviruses stably incorporate their genetic material and thus the cargo gene into the host genetic material. A disadvantage is that multiple cargo gene copies can be inserted into the genome. Adenoviruses and adeno-associated viruses (AAV) leave behind a so-called episome in the host cell, which contains the genetic information including the cargo gene. Episomes are not passed on during cell division and are therefore lost after some time. *Glybera*—the first gene therapy in the Western world—was based on AAV, as was *Luxturna,* a therapy for hereditary blindness approved in Europe since November 2018. Often, gene therapy involves removing stem cells from patients' bodies, which are then infected with the viruses in the lab and then reinserted into the affected individuals. The treatment of a single person can cost up to € 500,000

approval from the university's ethics committee to carry out the first gene therapy in Germany. In the 1990s, reservations were high because the trauma of the eugenic misuse of genetic knowledge during the Nazi era still overshadowed its use in humans. The therapy was directed against **cancer cells** of a patient. For this purpose, skin cells were taken from him and genetically modified to produce a natural messenger of the immune system, interleukin-2. The cells thus reconstructed were mixed with the patient's tumour cells, inactivated and re-injected into the patient. Interleukin-2, which is produced by the skin cells for about

3 weeks, is then thought to stimulate the immune system to attack and break down the tumour cells. Today, as of December 2018, there are over 2930 gene therapy trials registered worldwide, 656 of which are in Europe and 102 in Germany. The number is likely to rise sharply in the coming years, as the CRISPR/Cas system in particular is a promising and adaptable new method (Sect. 5.1).

The basic aim of **gene therapy** is the introduction of genetic information (cargo gene) into the genome of body tissue cells (somatic cells) for the treatment or prevention of disease. The genetic information can be used to replace, switch off or repair defective genes. In humans, one uses the ability of viruses, which naturally introduce genetic information in the form of DNA or RNA into infected cells (host cells) (Fig. 5.7) [42]. They are referred to as viral vectors. For this purpose, the cargo gene is inserted into the viral genome. In parallel, in most cases, tissue is removed from the patients and then infected with the virus in the laboratory and reimplanted. **Retroviruses** stably integrate their genome into the genetic material of the host cell, while **adenoviruses** place it as a so-called episome. This is degraded over time. Of course, viruses are used whose activity can be well controlled. There are also non-viral systems in which naked DNA is injected, but their efficiency is still low. In animals, numerous other methods are available which are (still?) prohibited in humans for ethical reasons (Sect. 3.5).

Only gene therapy on **somatic cells** is permitted. These are all cells that are not used for reproduction in humans. As early as 1985, a group of experts appointed by the German *Federal Ministry of Justice and Research* came to the conclusion in its final report that the insertion of genetic material into somatic cells is basically no different from organ transplantation in terms of its ethical dimension. The *German Medical Association* came to the same conclusion in its guidelines on gene therapy published in 1989.

In somatic gene therapy, two fundamentally different methods can be distinguished, the *ex vivo* and the *in vivo method*.

In the ***ex vivo* method**, also called the *in vitro method*, diseased cells are taken from patients and cultured (grown) outside the body. Cells from this cell culture are used to make the genetic modification. The genetically modified cell is first multiplied and then transferred back into the diseased individuals. However, not all cell types are suitable for cultivation outside the body. In these cases, the second method must be used, the ***in vivo* method**. For this purpose, the patients are usually injected with viruses that contain the correct genetic material. The viruses have therefore been genetically modified beforehand. In the body, these viruses ensure that the correct genetic material is introduced into the human cells, where it becomes active. Attempts are being made to optimise the viruses as gene shuttles so that they specifically transfer the desired genetic information to the desired tissue cells. In addition, the viruses must not lead to an excessive burden on the immune system. This happened in 1999 when the 18-year-old US American Jesse Gelsinger died during gene therapy. He suffered from a severe metabolic disease that was to be treated with gene therapy. However, his immune system was already compromised at the beginning of the therapy and he also received an excessive amount of viruses. As a result, he died of multiple organ failure. In principle, somatic gene therapy aims to repair defective body cells.

The situation is completely different for **germ cells**, that is egg and sperm cells, from which the embryo develops after fertilisation. These so-called germline cells are subject to the German **Embryo Protection Act**, which also defines them very precisely as all cells leading in a cell line from the fertilised ovum to the egg and sperm cells of the human being resulting from it, furthermore the ovum from the insertion or penetration of the sperm cell to the fertilisation completed with the nuclear fusion.

Furthermore, the law clearly states that *artificial modification of the genetic information of human germ line cells* is prohibited. Embryos may be created in Germany solely with the aim of bringing about a pregnancy, for example in the context of artificial insemination.

Regardless of the German view on the handling of germ cells, research on the germ line is already being carried out officially in other countries. Scientists in the team of English developmental biologist Kathy Niakan at the *Francis Crick Institute* in London have been allowed to use the CRISPR/Cas system on **human embryos** since February 2016. The research is clearly not about researchers growing babies in the lab. It is about basic research on early division stages of fertilized eggs. One of the research questions is the occurrence of side effects after genetic treatment of the cells. Unlike in Germany, where the Embryo Protection Act makes it clear that the *fertilised, viable human egg* cell is life worthy of protection from *the time of nuclear fusion*, in England this only applies from the 14th day after fertilisation. Therefore, embryos in Niakan's research are killed after 7 days. The **14-day rule** applies to embryo researchers in countries such as Australia, Canada, the US, Denmark, Sweden and the UK. The time limit for using artificially created embryos for scientific purposes no longer than 14 days after fertilisation is based on developmental biology. Around the 14th day, the so-called **primitive streak**, which is a first sign of a developing nervous system, appears (Fig. 4.11). Moreover, after the 14th day, it is impossible for the embryo to divide to form twins. Prior to this, the formation of identical twins would be possible, which can be considered a lack of **individuality**. Accordingly, in the period before the formation of this individuality, there is still no dignity, but only respect for the species of man. Man as a living being is regarded as something different from man as a person. Therefore, as early as 1984, English scientists proposed to allow embryo research until the 14th day. Since it became possible in 2016 to keep human embryos alive outside the womb for longer than 14 days after artificial insemination, this rule has been hotly debated. For **Muslims**, by the way, life begins with the entrance of the soul. This happens after 120 days, before which abortion, for example, is permitted.

A special form of germline therapy is *mitochondrial replacement therapy* (MRT). It results in so-called **three-parent babies**. It is used when there is a genetic defect in the hereditary material of the power plants of the cells, the mitochondria (Fig. 2.3), which is the case in about one in 5000 births. One of these defects is **Leigh Syndrome**. This is a severe metabolic disease with a life expectancy of a few years. There are 36 genes encoded in the mitochondrial genome that are only passed on from the mother through the egg. In the corresponding germ line therapy, the defective mitochondria are replaced by mitochondria from an egg cell donor (Fig. 5.8).

This means that less than 0.00001% of the genetic information is replaced by the natural genetic information of the mitochondria of the donor mother. The *British House of Commons* decided in 2015 to allow this intervention in the human germ line under certain conditions. U.S. regulators have ruled the method safe only for male embryos, but have not yet approved it. Worldwide, three three-parent babies have been born as of April 2019, the first in Mexico in 2016, one in Ukraine and one in Spain [43, 44]. Two pairs of parents in the UK received approval to use the procedure in 2018.

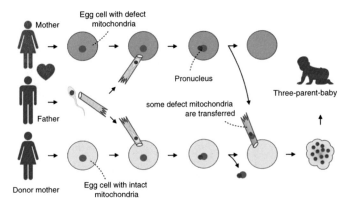

Fig. 5.8 In the case of three-parent babies, it has been officially permitted to intervene in the germ line of humans in England since 2015. In this way, mitochondria with a defective genome are replaced by intact mitochondria (not shown) from a donor mother. The pronucleus consists of the cell nuclei with the chromosomes of the parents

It is becoming clear that intervention in the human germ line for research purposes is in full swing. Perhaps the most important ethical question in the near future will be how we deal with the knowledge gained from genetic engineering (Sect. 8.2) and the possibilities offered by genetic engineering—and this against very different socio-cultural backgrounds. If a serious hereditary disease cannot be treated but can be repaired, do doctors not have to help? How should abortion be weighed against gene therapy intervention in the germ line? These are just two of many questions that need to be answered.

References

1. Cribbs AP, Perera SMW (2017) Science and Bioethics of CRISPR-Cas9 Gene Editing: An Analysis Towards Separating Facts and Fiction. Yale J Biol Med 90: 625–634

2. Hardt A (2019) Technikfolgenabschätzung des CRISPR/
   Cas-Systems. De Gruyter, Berlin
3. Aslan SE, Beck B, Deuring S, et al (2018) Genom-Editierung
   in der Humanmedizin: Ethische und rechtliche Aspekte von
   Keimbahneingriffen beim Menschen. In: CfB-Drucksache 4.
   Visited 23.04.2019: uni-muenster.de/imperia/md/content/
   bioethik/cfb_drucksache_4_2018_genom_editier-
   ung_13_06_final.pdf
4. Jinek M, Chylinski K, Fonfara I, et al (2012) A Programmable
   Dual-RNA-Guided DNA Endonuclease in Adaptive
   Bacterial Immunity. Science 337: 816–821. doi:https://doi.
   org/10.1126/science.1225829
5. Cong L, Ran FA, Cox D, et al (2013) Multiplex genome
   engineering using CRISPR/Cas systems. Science 339:
   819–823. doi:https://doi.org/10.1126/science.1231143
6. Shinagawa H, Makino K, et al (1987) Nucleotide sequence
   of the iap gene, responsible for alkaline phosphatase isozyme
   conversion in *Escherichia coli,* and identification of the gene
   product. J Bacteriol 169: 5429–5433. doi:https://doi.
   org/10.1128/jb.169.12.5429-5433.1987
7. Mojica FJ, Juez G, Rodríguez-Valera F (1993) Transcription at
   different salinities of *Haloferax mediterranei* sequences adjacent
   to partially modified PstI sites. Mol Microbiol 9: 613–621.
   doi:https://doi.org/10.1111/j.1365-2958.1993.tb01721.x
8. Mojica FJ, Ferrer C, Juez G, Rodríguez-Valera F (1995) Long
   stretches of short tandem repeats are present in the largest
   replicons of the Archaea *Haloferax mediterranei* and *Haloferax
   volcanii* and could be involved in replicon partitioning. Mol
   Microbiol 17: 85–93. doi:https://doi.org/10.1111/j.
   1365-2958.1995.mmi_17010085.x
9. Mojica FJM, Díez-Villaseñor C, García-Martínez J, Soria E
   (2005) Intervening sequences of regularly spaced prokaryotic
   repeats derive from foreign genetic elements. J Mol Evol 60:
   174–182. doi:https://doi.org/10.1007/s00239-004-0046-3
10. Jansen R, van Embden JDA, Gaastra W, Schouls LM (2002)
    Identification of genes that are associated with DNA repeats

in prokaryotes. Mol Microbiol 43: 1565–1575. doi:https://doi.org/10.1046/j.1365-2958.2002.02839.x

11. Makarova KS, Grishin NV, Shabalina SA, et al (2006) A putative RNA-interference-based immune system in prokaryotes: computational analysis of the predicted enzymatic machinery, functional analogies with eukaryotic RNAi, and hypothetical mechanisms of action. Biol Direct 1: 7. doi:https://doi.org/10.1186/1745-6150-1-7

12. García-Martínez J, Maldonado RD, Guzmán NM, Mojica FJM (2018) The CRISPR conundrum: evolve and maybe die, or survive and risk stagnation. Microb Cell 5: 262–268. doi:https://doi.org/10.15698/mic2018.06.634

13. Barrangou R, Fremaux C, Deveau H, et al (2007) CRISPR provides acquired resistance against viruses in prokaryotes. Science 315: 1709–1712. doi:https://doi.org/10.1126/science.1138140

14. Marraffni LA (2015) CRISPR-Cas immunity in prokaryotes. Nature 526: 55–61. doi:https://doi.org/10.1038/nature15386

15. Schmidt F, Cherepkova MY, Platt RJ (2018) Transcriptional recording by CRISPR spacer acquisition from RNA. Nature 562: 380–385. doi:https://doi.org/10.1038/s41586-018-0569-1

16. Pawluk A, Davidson AR, Maxwell KL (2018) Anti-CRISPR: discovery, mechanism and function. Nat Rev Microbiol 16: 12–17. doi:https://doi.org/10.1038/nrmicro.2017.120

17. Wagner DL, Amini L, Wendering DJ, et al (2018) High prevalence of *Streptococcus pyogenes* Cas9-reactive T cells within the adult human population. Nat Med 25: 242–248. doi:https://doi.org/10.1038/s41591-018-0204-6

18. Liang P, Xu Y, Zhang X, et al (2015) CRISPR/Cas9-mediated gene editing in human tripronuclear zygotes. Protein Cell 6: 363–372. doi:https://doi.org/10.1007/s13238-015-0153-5

19. Society and Ethics Research Wellcome Genome Campus (2018) International Summit on Human Genome Editing—He Jiankui presentation and Q&A. In: YouTube. Visited 03.12.2018: youtu.be/tLZufCrjrN0

20. Hütter G, Nowak D, Mossner M, et al (2009) Long-Term Control of HIV by CCR5Δ32/Δ32 Stem-Cell

Transplantation. N Engl J Med 360: 692–698. doi:https://doi.org/10.1056/nejmoa0802905

21. Allers K, Hütter G, Hofmann J, et al (2011) Evidence for the cure of HIV infection by CCR5Δ32/Δ32 stem cell transplantation. Blood 117: 2791–2799. doi:https://doi.org/10.1182/blood-2010-09-309591

22. Gupta RK, Abdul-Jawad S, McCoy LE, et al (2019) HIV-1 remission following CCR5Δ32/Δ32 haematopoietic stemcell transplantation. Nature 568: 244–248. doi:https://doi.org/10.1038/s41586-019-1027-4

23. Keele BF (2006) Chimpanzee Reservoirs of Pandemic and Nonpandemic HIV-1. Science 313: 523–526. doi:https://doi.org/10.1126/science.1126531

24. Novembre J, Galvani AP, Slatkin M (2005) The Geographic Spread of the CCR5Δ32 HIV-Resistance Allele. PLoS Biol 3: e339. doi:https://doi.org/10.1371/journal.pbio.0030339

25. Galvani AP, Slatkin M (2003) Evaluating plague and smallpox as historical selective pressures for the CCR5-Δ32 HIV-resistance allele. Proc Natl Acad Sci USA 100: 15276–15279. doi:https://doi.org/10.1073/pnas.2435085100

26. Falcon A, Cuevas MT, Rodriguez-Frandsen A, et al (2015) CCR5 deficiency predisposes to fatal outcome in influenza virus infection. J Gen Virol 96: 2074–2078. doi:https://doi.org/10.1099/vir.0.000165

27. Joy MT, Ben Assayag E, Shabashov-Stone D, et al (2019) CCR5 Is a Therapeutic Target for Recovery after Stroke and Traumatic Brain Injury. Cell 176: 1143–1157.e13. doi:https://doi.org/10.1016/j.cell.2019.01.044

28. Zhou M, Greenhill S, Huang S, et al (2016) CCR5 is a suppressor for cortical plasticity and hippocampal learning and memory. eLife 5: 338. doi:https://doi.org/10.7554/elife.20985

29. Ledford H (2015) Where in the world could the first CRISPR baby be born? Nature 526: 310–311. doi:https://doi.org/10.1038/526310a

30. Ishii T (2017) Germ line genome editing in clinics: the approaches, objectives and global society. Briefings Funct

Genomics 16: 46–56. doi:https://doi.org/10.1093/bfgp/elv053

31. Jiankui H, Ferrell R, Yuanlin C, et al (2018) Draft Ethical Principles for Therapeutic Assisted Reproductive Technologies. CRISPR J. doi:https://doi.org/10.1089/crispr.2018.0051.retract (this paper has been retracted)

32. Cheng Y (2019) Brave new world with Chinese characteristics. In: Bulletin of the Atomic Scientists. Visited 23.02.2019: thebulletin.org/2019/01/brave-new-world-with-chinese-characteristics/

33. Yang X (2019) Weltmacht: Ob in China … Die Zeit, Ausgabe 16, Seite 3

34. Krimsky S (2019) Ten ways in which He Jiankui violated ethics. Nat Biotechnol 37: 19–20. doi:https://doi.org/10.1038/nbt.4337

35. Schöne-Seifert B (2019) "Russisches Roulette" in der Genforschung am Menschen? Ethik Med 362: 1–5. doi:https://doi.org/10.1007/s00481-018-00516-z

36. Fischer J (2018) Der Abstieg des Westens: Europa in der neuen Weltordnung des 21. Jahrhunderts. Kiepenheuer & Witsch, Köln

37. Liu Z, Cai Y, Wang Y, et al (2018) Cloning of Macaque Monkeys by Somatic Cell Nuclear Transfer. Cell 172: 881–887.e7. doi:https://doi.org/10.1016/j.cell.2018.01.020

38. Liu Z, Cai Y, Liao Z, et al (2019) Cloning of a gene-edited macaque monkey by somatic cell nuclear transfer. Natl Sci Rev 6: 101–108. doi:https://doi.org/10.1093/nsr/nwz003

39. Al-Balas QA, Dajani R, Al-Delaimy WK (2019) Traditional Islamic approach can enrich CRISPR twins debate. Nature 566: 455. doi:https://doi.org/10.1038/d41586-019-00665-1

40. Lander ES, Baylis F, Zhang F, et al (2019) Adopt a moratorium on heritable genome editing. Nature 567: 165–168. doi:https://doi.org/10.1038/d41586-019-00726-5

41. Salganik M, Hirsch ML, Samulski RJ (2015) Adeno-associated Virus as a Mammalian DNA Vector. In: Craig, Chandler, Gellert, et al (Hrsg) Mobile DNA III. American

Society of Microbiology, S 829–851. doi:https://doi.org/10.1128/microbiolspec.MDNA3-0052-2014

42. Kay MA (2011) State-of-the-art gene-based therapies: The road ahead. Nat Rev Genet 12: 316–328. doi:https://doi.org/10.1038/nrg2971

43. Zhang J, Zhuang G, Zeng Y, et al (2016) Pregnancy derived from human zygote pronuclear transfer in a patient who had arrested embryos after IVF. Reprod BioMed Online 33: 529–533. doi:https://doi.org/10.1016/j.rbmo.2016.07.008

44. Reardon S (2016) "Three-parent baby" laim raises hopes–and ethical concerns. Nature News. doi:https://doi.org/10.1038/nature.2016.20698

Further Reading

Adli M (2018) The CRISPR tool kit for genome editing and beyond. Nat Commun 9: 1911. doi:https://doi.org/10.1038/s41467-018-04252-2

Chandrasegaran S, Carroll D (2016) Origins of Programmable Nucleases for Genome Engineering. J Mol Biol 428: 963–989. doi:https://doi.org/10.1016/j.jmb.2015.10.014

Donohoue PD, Barrangou R, May AP (2018) Advances in Industrial Biotechnology Using CRISPR-Cas Systems. Trends Biotechnol 36: 134–146. doi:https://doi.org/10.1016/j.tibtech.2017.07.007

Lander ES (2016) The Heroes of CRISPR. Cell 164: 18–28. doi:https://doi.org/10.1016/j.cell.2015.12.041

Rommelfanger KS, Wolpe PR, Drafting T, Drafting and Reviewing Delegates of the BEINGS Working Groups (2017) Ethical principles for the use of human cellular biotechnologies. Nat Biotechnol 35: 1050–1058. doi:https://doi.org/10.1038/nbt.4007

6
Writing Genetic Material

While in the previous sections we learned about very concrete interventions in the genetic material of organisms, things are now getting a little more utopian—but only a little. The latest methods of genetic engineering involve not only the precise modification of the genetic material without leaving traces, but also the completely chemical synthesis of genetic information. The DNA molecule can thus be synthesized in a test tube. This is nothing new: Old **DNA synthesis devices** can be obtained on *eBay* for relatively little money (Fig. 6.1). However, the technology is becoming more sophisticated. The currently available phosphoramidite-based chemical synthesis can generate fragments around 250 nucleotides long. With the development of newer methods, fragment lengths are becoming much larger [1].

The bacteria-infecting virus (phage) **ϕX174** was the first genome to be sequenced (by Frederik Sanger) in 1977— and a quarter of a century later one of the first to be fully synthesised [2, 3]. A pioneer of genome research, Craig Venter, was involved in this work. Before that, the German-born biochemist Eckard Wimmer, who was doing research in the USA, had succeeded in synthetically generating an

© Springer-Verlag GmbH Germany, part of Springer Nature 2022
R. Wünschiers, *Genes, Genomes and Society*,
https://doi.org/10.1007/978-3-662-64081-4_6

Fig. 6.1 Still available: an apparatus for the chemical synthesis of short DNA fragments on *eBay*

infectious **poliovirus**. We have thus recently become able to write genetic information. In 2010, the first bacterium was created whose genetic material was not generated by the bacterium itself, but in a test tube [4]. Scientists at the *J Craig Venter Institute* said they put 15 years of work and $40 million into the project. The resulting bacterium is called *Mycoplasma* mycoides strain JCVI-syn1.0 (Fig. 6.4).

6.1   Fabrication of Life

It is true that science is still far from being able to produce complete living beings from a mixture of chemicals. However, the idea and the fascination associated with it are as old as the hills, as the extract from Francis Bacon's *New Atlantis* shows (Chap. 3). An impressive attempt to recreate life was made by the French engineer and inventor Jacques De Vaucanson in the mid-eighteenth century (Fig. 6.2) [5]. His **duck** was composed of several hundred movable structural elements (one wing alone consisted of over 400 parts) and could *Waddle. Drink. Eat. Shit*, as the German writer

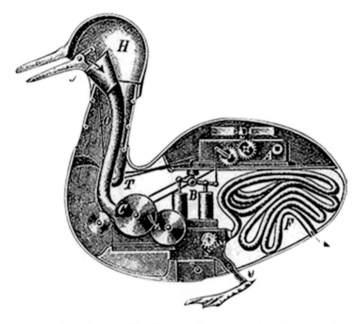

Fig. 6.2 Fabricating life. The mechanical duck of Jacques De Vaucanson

Günter Kunert wrote in a short story about the duck [6]. Unfortunately, no specimen survived, but the French philosopher Denis Diderot, a contemporary of De Vaucanson, saw it and included it in his 1751 encyclopedia under the entry *automaton*. A replica can still be admired today in the *Musée des Automates de Grenoble Rêves mécaniques* in Grenoble, France. The imitation of nature went so far that the duck even had an artificial digestive system. Grains that it picked up were digested and excreted by a chemical cocktail in its artificial intestine.

Current attempts to recreate living systems are no different in idea from earlier ventures. And when I write systems, it already becomes clear that the underlying mindset is that of an engineer. A living thing is considered a biological system. The science of regulating this system was christened **cybernetics** in the mid-nineteenth century. Due to the possibilities of measurement and analysis of biological systems in high throughput, cybernetics became **systems biology**. It tries to describe the biological system as comprehensively as possible. The consequent subsequent step is to use this systemic knowledge not only to control life processes but also to design and extend them from scratch. In Sect. 2.2 we have already learned about the genetic code, according to which three nucleotides each form a codon (triplet), which in turn code for one of the 20 amino acids. In some organisms, the stop codon TGA can also code for a 21st amino acid, **selenocysteine**, or a 22nd, **pyrrolysine**. In humans, 25 proteins are currently known to contain selenocystin, so-called selenoproteins, which counteract oxidative stress in particular and appear to play an important role in our nervous system [7, 8]. What could be more logical than to encode a 23rd, 24th... amino acid to extend the code. Theoretically, a triplet of four nucleotides can encode 64 amino acids ($4 \times 4 \times 4$). However, some amino acids have several different triplets. For example, the amino acid serine is encoded by the four triplets TCA, TCC, TCG and TCT. The simplest way is to assign a new amino acid to a triplet. This is complex because numerous molecular components in the cell are involved in the translation of a triplet into an amino acid. However, it has already been possible to genetically engineer corresponding microorganisms [9]. In this way, amino acids can be incorporated into proteins in a way that does not occur in nature and that is almost impossible to produce in the laboratory using chemical synthesis. This not only helps in the study of the function of biological systems but it is also hoped that new drugs can be developed in this way.

Scientists from Cambridge in the UK are trying to add a nucleotide to the molecular translation apparatus in the cell

[10]. A triplet becomes a **quadruplet**, which can theoretically encode 4 × 4 × 4 = 256 amino acids. This is an even deeper intervention in the translational apparatus. In early 2019, a team of researchers from the USA succeeded in adding four additional **synthetic nucleotides** to the genetic code [11]. In addition to the natural nucleotides A, T, C and G (Sect. 2.1), B, P, S and Z have been added, whose molecule names I will spare you at this point. B pairs with S and P with Z. The researchers have named their synthetic DNA *hachimoji-DNA*, which means eight letters in Japanese. In this way, not only can more amino acids be encoded but it is also possible to build a biological system that is no longer compatible with natural systems (orthogonality, Sect. 6.2). The application of this "extreme genetic engineering" should thus contribute to safety, as outcrossing is no longer possible. In addition, the *hachimoji-DNA* should be able to function as an information store. Research has long been underway to find a medium that can be used to store information for a long time. Anyone who still has old CDs will be surprised when they can no longer be played. On the other hand, it has already been possible to read 50,000-year-old DNA molecules (Sect. 4.2). The examples described show that bioengineering has finally been born. This new research direction is called **synthetic biology**.

6.2 Synthetic Biology

Synthetic biology is attracting attention with headlines such as *We are playing God* in popular magazines. They describe how scientists are trying to specifically modify microorganisms in order to give them functions such as the synthesis of biodiesel or the binding of atmospheric carbon

182   R. Wünschiers

dioxide. Isn't that genetic engineering? Indeed, genetic engineering has been dealing with similar issues for a long time. Synthetic biology, however, goes further by elevating biotechnology and genetic engineering to a true art of engineering that works with standardised components or **DNA modules** (Fig. 6.3).

As in engineering, the result of the recombination (synthesis) of such components should be predictable or simulatable (see *New Atlantis* in Chap. 3). Trial and error are to be replaced by design. Synthetic biology, as a modern subdiscipline of the life sciences, is currently the culmination of a series of genetic engineering developments of recent decades.

At the outset, I would like to avoid the misunderstanding that synthetic biology is synthetic evolutionary theory.

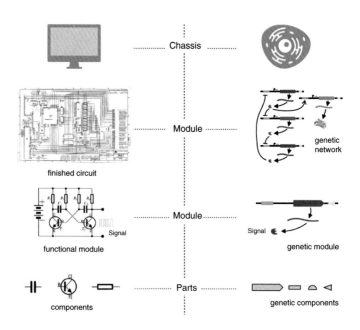

Fig. 6.3 Concept of synthetic biology. As in electronics, standardized components (parts) are to be assembled into functional components (modules) and installed in chassis

While the latter is an extension of Darwin's theory of evolution to include molecular biological findings, synthetic biology involves a new research and application field within the life sciences. Another common misunderstanding concerns the term **synthetic**: synthesis is to be understood here in the sense of bringing together—DNA modules with defined functions are brought together and something new is created.

The ideas behind synthetic biology are not new. As early as 1911, the biologist Jacques Loeb (1859–1924), who was born in Mayen, Germany but emigrated to the USA, argued in a lecture in Hamburg that the way to understand the nature of life was to produce life in the laboratory. His views were published in the magazine *Popular Science Monthly* in 1912 and were published as a book entitled *The Mechanistic Conception of Life* later that year [12]. The greatest influence on the public perception of synthetic biology to date has been the media-staged publication of the research results of the *J Craig Venter Institute* in the USA, on 20 May 2010 as a press release, press conference and scientific publication. It was announced that

> the first self-replicating species that we have had on the planet whose parent is a computer [...] the first species that has its own website encoded in its genetic code

was created. The synthesis seems simple: *building the chromosomes from four bottles of chemicals*. The non-profit *J-Craig Venter Institute* triggered a worldwide media response, ranging from serene awareness to hysterical Frankenstein reports. The Vatican calmly received the news of the first synthetic species called **Mycoplasma mycoides JCVI-syn1.0** and acknowledged the scientific achievement. This is not without a certain irony, since most press releases, at least in Europe, accused the scientists of playing God.

But what has Craig Venter's research group actually achieved (Fig. 6.4)? First, an altered genome sequence of the bacterium *Mycoplasma mycoides* was synthesized. The changes compared to the approximately one million nucleotide long sequence template mainly concern the integration of so-called **watermarks**, which make the synthetic genome clearly distinguishable from the original. A watermark contains, for example, the URL to a website with information about the bacterium. The actual synthesis of the genome took place in several steps. First, the chemical synthesis of 1078times 1080 nucleotide-long DNA fragments

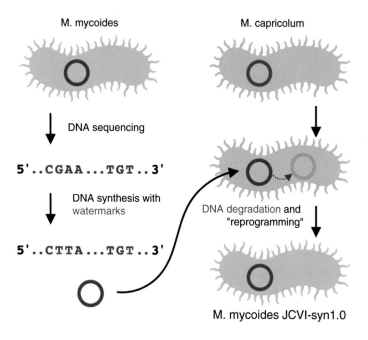

Fig. 6.4 Complete chemical synthesis of the chromosome of the bacterium *Mycoplasma mycoides* and transfer of the DNA into the bacterium *Mycoplasma capricolum*. In this way, one organism is reprogrammed into the other

was commissioned from DNA synthesis companies—including the German company *GeneArt* from Regensburg. Since the sequence ends of the ordered fragments partially overlapped, they could be fused (ligated) to 109 times 10,080 nucleotide long fragments. In a further step, ligation was carried out to form fragments elven times 100,000 nucleotides long and finally the final fusion of the 11 fragments to form a circular chromosome 1,077,947 base pairs long. However, the aforementioned fragment fusions were not performed *in vitro* (in a test tube), but were carried out in yeast cells. The described **genome synthesis** thus mainly uses biological functions of yeast. Only the initial synthesis of the short fragments was carried out chemically. Also, the introduction of the synthetic *Mycoplasma mycoides* genome into *Mycoplasma capricolum* was performed by a standard cell biological method, protoplast fusion (Fig. 3.4). Soberly considered, the scientific success of this approximately 15-year research project is reduced to the combination of old methods and the elimination of new problems resulting from them—as well as great public relations. Since both the yeast and the target protoplast had to be functional and completely present, there can be no question of a new synthesis of life.

Alongside Craig Venter, the US geneticist George Church is one of the great pioneers in the field of synthetic biology. He is more interested in the targeted modification of the genetic information of an existing genome than in synthesising a new one. To this end, his research group also works with the CRISPR/Cas system. If one wants to edit many sites in the genome at the same time, a major problem arises: the DNA breaks down into many parts (fragmented). If the gene editing systems cuts at one position, the DNA is reassembled by the cell's own repair mechanisms (Fig. 5.2). However, if, for example, the gene editing cuts at a hundred positions in parallel, then too many chromosome fragments are present at the same time and the repair mechanism fails. George Church's team recently unveiled an advanced nuclease that allowed them to introduce over 13,000 changes simultaneously on human cells [13]. Back in 2017, scientists with the participation of George Church were able to use a similar method to cut retroviruses (*porcine endogenous retroviruses,* PERV) from the genome of pigs at 25 positions and give birth to correspondingly altered piglets, delivered after *in-vitro* fertilization of foster sows [14]. This research on pigs is of great importance in connection with **xenotransplantation**, in which attempts are made to transplant organs from donor pigs into humans [15]. This is because, in addition to the immune defence against organ tissue from animals, the possible transmission of viruses is one of the main problems.

But George Church thinks even further: he wants to revive extinct animals (Fig. 6.5) [16]. The first extinct animal to be reborn as a clone was the Pyrenean ibex, which became extinct in 2000 [17]. However, the clone lived only a

Fig. 6.5 An *ostrichosaurus*—there is no such thing as a strawberry that big. Birds are still living dinosaurs. Some scientists think it is possible to change the genetic material and accordingly such a bird as the ostrich into dinosaur-like creatures. (Source: The drawing of the strawberry gardener is by Inga-Lisa Burmester)

few minutes. But what about creatures that have been extinct for millions of years? Can we bring them back to life as in Michael Crichton's *Dino Park* and the film adaptation *Jurassic Park* respectively [18]? Church believes we can. Starting with the closest living relatives of the dinosaurs, the birds, the mutations that separate the two classes would have to be reversed step by step. But this would require knowing the genetic differences. And indeed, several teams of scientists are trying to predict them [19]. The differences between Neanderthal man and modern man, on the other hand, are known ...

What I cannot create, I cannot understand, the US physicist and Nobel Prize winner Richard Feynman once said. It is only a small step from the mathematical modelling of metabolic pathways and the prediction of their behaviour with a computer to the validation or use in a living organism—if the methodology for the targeted generation of the designed desired organism exists. This outlines an important area of research in synthetic biology: the design of a metabolic pathway or even an organism and its incarnation. The results of the modelling provide the goal. This part of synthetic biology is called **metabolic design** and is a direct continuation of genetic engineering. Examples are the biosynthesis of the malaria drug artemisinic acid (see later) or of biodiesel with specifically designed microorganisms.

Why is this more than genetic engineering? The essential difference lies in the procedure. All biological functional elements are available as so-called **parts** (functional biological components) together with a detailed description in an electronic database and as a DNA sequence in a plasmid. Scientists can now select suitable components on the basis of the description and then physically order them.

The plasmids into which the DNA sequences of the parts are cloned have a strictly defined structure that allows several parts to be arranged in sequence to form a gene construct (**module**) using standard molecular biology methods. This can then be used to transform a target organism. Synthetic biology is intended to introduce an engineering-like approach to genetic engineering. Very well characterized components (here parts) are brought together to perform a predefined task in the target organism. Biologists become constructors. This standardisation of parts can be compared to the introduction of ISO standards in biotechnology. Biological components can now be assembled, by humans or robots, using standard methods, regardless of their source.

Let us take a look at the flagship project of synthetic biology and metabolic engineering: the antimalarial drug **artemisinin**. Malaria, which occurs mainly in parts of the world with tropical or subtropical climates, is caused by a single-celled parasite belonging to the genus *Plasmodium* and transmitted by mosquitoes. According to estimates by the *World Health Organization* (WHO), around 200 million people fall ill every year, of whom around one million die. Various preparations are available to combat the disease. However, resistance to the pathogen is increasingly developing [20]. The WHO therefore recommends an artemisinin combination preparation (trade name *Eurartesim*) with the active ingredients dihydroartemisinin and piperaquine phosphate. The active substance dihydroartemisinin is extracted from the mugwort plant *Artemisia annua.* One problem is that the amount produced by the plants cannot meet current world demand [21]. Using methods of synthetic biology, especially metabolic design, it has become possible to synthesize the precursor of this substance in large quantities in microorganisms and keep the cost low enough to keep it affordable for patients in developing countries. The research was funded by the *Bill & Melinda Gates Foundation* and the production process was licensed by *Sanofi* in 2008. This is the first drug to be produced on an industrial scale using synthetic biology [22].

Biotechnological production was achieved by modifying and expanding the metabolism of the bacterium *Escherichia coli* and the yeast *Saccharomyces cerevisiae* [23, 24]. For the production in yeast as a production organism, only four steps need to be added in order for artemisinic acid to be produced. By up- and downregulation of certain genes in this newly constructed metabolic pathway, additional improvements in the conversion of the individual synthesis steps are achieved. Finally, the produced artemisinic acid is transported out of the cell and converted into the final drug. The comparison between the conventional production pathway using mugwort and recombinant production using bacteria or yeasts clearly shows the advantages: While extraction from the plant takes about a year and depends on climatic and environmental factors, synthesis by microorganisms takes place within about 4 weeks, independent of the environment.

Another branch of synthetic biology deals with the target organism, which is consequently called **chassis**. The goal is to design a chassis that interferes as little as possible with the gene constructs to be included. There are two elementary approaches to this. In the research area of **minimal cells**, starting from a bacterium with a genome as small as possible, one investigates how many genes are dispensable. *Mycoplasma genitalium* has the smallest known genome of a free-living bacterium with just under 500 protein-coding genes. Studies indicate that 387 protein-coding genes and 43 RNA-coding genes are essential. The chassis with just these genes was presented to the public as *Mycoplasma laboratorium* 2016 [25] and a patent application was filed. In contrast to this **top-down approach**, in which genes are removed step by step starting from an intact cell, is the **bottom-up approach** of **protocell re-search** (Fig. 6.6).

Based on lipid vesicles, attempts are being made to compartmentalize genetic and biochemical components and to induce them to grow and replicate. This research approach is closely related and interwoven with the search for suitable conditions under which the emergence of life could have taken place. It remains to be seen which chassis will prevail in the long run.

Synthetic biology also involves attempts to design micro-organisms that act as biological **sensors** and provide a defined response to specific environmental signals. For

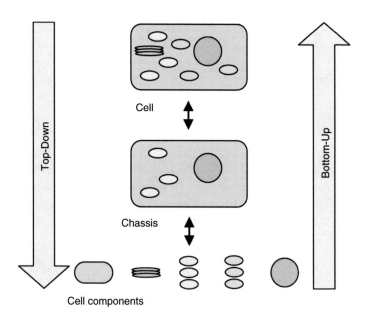

Fig. 6.6 Two different concepts of how cells can be constructed as chassis for genetic engineering applications. In the top-down approach, cells are progressively reduced. In contrast, in the bottom-up approach cells are created from scratch

example, there are attempts to express proteins in bacteria that have been optimised to detect the **explosive TNT** and trigger a reaction cascade in the bacterium when TNT is present. This could be designed to cause the bacteria to emit optical signals. The declared aim is to use such bacteria for the detection of land mines. It would also be conceivable to detect mixtures of substances with an organism and also to display concentration levels. This would require **complex circuits**, another branch of synthetic biology. The ultimate goal is logical signal processing as known from computers. The foundations for this have already been laid. The metaphor of programming a cell is thus raised to a new level, from programming the DNA code to programming information processing through genetic control circuits. A major success was indeed achieved in 2000 with the targeted development and simulation of a circuit on the computer and its implementation in a cell [26]. The company *Microsoft Research* has already developed a programming language and environment specifically tailored to the demands of synthetic biology called Visual GEC, where GEC stands for *genetic engineering of living cells* [27].

Regardless of whether a chassis or a complex target organism is being worked with: Interactions between the basic functions of the chassis and the introduced modules must be prevented. The subfield of **orthogonal systems** explicitly addresses this problem. For example, by using an unnatural genetic code with an adapted gene expression apparatus, the synthetic transcription and translation apparatus can be separated from that of the target organism.

The geneticist George Church thinks even further: The proteins of all living organisms are composed of amino acids that have a specific spatial atomic arrangement, the so-called L-amino acids. For every L-amino acid there is a mirror-inverted D-variant, a so-called stereoisomer. Only

L-stereoisomers are found in all living organisms. Like a screw, DNA has a direction of rotation, too. Church has proposed trying to create a bacterium composed only of R-amino acids. His belief is that the L and R organisms would be incompatible and thus no exchange of genetic information could occur—a perfectly orthogonal system. Church's idea, by the way, is not to be confused with *situs inversus* (Latin for inverted position). This is an externally undetectable anatomical feature that occurs in about one in 20,000 people, in which the organs are mirror images of each other. The heart sits on the right and so on. The heritability of *situs inversus* was first described in snails [28].

Orthogonal systems are also frequently mentioned in the context of **biological safety** (biosafety). This subfield of synthetic biology deals with the risks of using synthetic microorganisms and overlaps almost completely with safety research in genetic engineering. Overall, biosafety measures serve to protect employees, the population and the environment from dangerous organisms and biological agents.

Let us do a thought experiment. Let us assume that we can kill a modified organism with an efficiency of 99.9999%. In one millilitre of a cell suspension containing ten million bacterial cells, there is then statistically still one cell that could potentially escape non-deactivated from the laboratory or production plant. One safety measure would be the use of so-called auxothrophic organisms. These rely on absorbing essential nutrients from the environment. The goal is to create organisms that survive only by adding special substances not found in nature. Induced cell death is another method of preventing laboratory and production strains from surviving in the environment. This involves introducing metabolic pathways into the organism that lead to the formation of cellular toxins. These metabolic pathways could be switched on by light, for example.

A greater danger, however, is posed by the criminal or terrorist **misuse** (*dual-use problem*) of the possibilities of synthetic biology—a topic addressed by the area of **biosecurity**. In this context, the publication of the relatively simple synthesis of the poliovirus and the **Spanish flu** virus alarm [29]. The latter caused the death of about 50 million people in 1918. Remarkably, the publication on the poliovirus was extended by a note shortly before publication:

> Note added in proof: [...] The fundamental purpose of this work was to provide information critical to protect public health and to develop measures effective against future influenza pandemics.

The synthesis of both viruses was carried out against the background of understanding the molecular mechanisms of infection and extremely high pathogenicity. The public accessibility of the genome sequences of extremely pathogenic viruses, such as the **Ebola virus** (Fig. 6.7), combined with the possibility of ordering DNA molecules tailor-made as commodities, shows the explosive nature of what is possibly the simplest subfield of synthetic biology, **DNA synthesis**.

DNA synthesis companies have therefore agreed to always check customer orders for sequence similarities with known pathogens—but this cannot prevent criminal forces from ordering DNA synthesis equipment on *eBay* (Fig. 6.1). Like any technology, synthetic biology also harbours opportunities and risks, the potential of each of which must be weighed up.

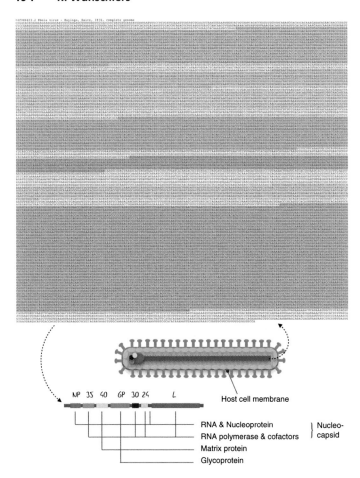

Fig. 6.7 The genome of the Ebola virus is only 18,959 nucleotides long. Since the genome consists of RNA, thymine (T) is replaced by uracil (U). The genome codes for seven proteins that make up the virus

References

1. Palluk S, Arlow DH, de Rond T, et al (2018) De novo DNA synthesis using polymerase-nucleotide conjugates. Nat Biotechnol 36: 645–650. doi:https://doi.org/10.1038/nbt.4173

2. Sanger F, Nicklen S, Coulson AR (1977) DNA sequencing with chain-terminating inhibitors. Proc Natl Acad Sci USA 74: 5463–5467. doi:https://doi.org/10.1073/pnas.74.12.5463

3. Smith HO, Hutchison CA, Pfannkoch C, Venter JC (2003) Generating a synthetic genome by whole genome assembly: φX174 bacteriophage from synthetic oligonucleotides. Proc Natl Acad Sci USA 100: 15440–15445. doi:https://doi.org/10.1073/pnas.2237126100

4. Gibson DG, Glass JI, Lartigue C, et al (2010) Creation of a bacterial cell controlled by a chemically synthesized genome. Science 329: 52–56. doi:https://doi.org/10.1126/science.1190719

5. Drux R (2017) "Eine höchst vollkommene Maschine": Von der poetischen Faszination einer mechanischen Ente im späten achtzehnten Jahrhundert. In: Zwischen Literatur und Naturwissenschaft. Walter de Gruyter Verlag, Berlin. S. 105–118. doi:https://doi.org/10.1515/9783110528114-005

6. Kunert G (1989) Tagträume in Berlin und andernorts. Fischer Taschenbuch Verlag, Frankfurt

7. Romagné F, Santesmasses D, White L, et al (2014) SelenoDB 2.0: Annotation of selenoprotein genes in animals and their genetic diversity in humans. Nucleic Acids Res 42: D437–D443. doi:https://doi.org/10.1093/nar/gkt1045

8. Reeves MA, Hoffmann PR (2009) The human selenoproteome: Recent insights into functions and regulation. Cell Mol Life Sci 66: 2457–2478. doi:https://doi.org/10.1007/s00018-009-0032-4

9. Xie J, Schultz PG (2006) A chemical toolkit for proteins—an expanded genetic code. Nat Rev Mol Cell Biol 7: 775–782. doi:https://doi.org/10.1038/nrm2005

10. Neumann H, Wang K, Davis L, et al (2010) Encoding multiple unnatural amino acids via evolution of a quadruplet-decoding ribosome. Nature 464: 441–444. doi:https://doi.org/10.1038/nature08817

11. Hoshika S, Leal NA, Kim M-J, et al (2019) Hachimoji DNA and RNA: A genetic system with eight building blocks. Science 363: 884–887. doi:https://doi.org/10.1126/science.aat0971

12. Loeb J (1912) The Mechanistic Conception of Life. The University of Chicago Press, Chicago, Illinois/USA

13. Smith CJ, Castanon O, Said K, et al (2019) Enabling large-scale genome editing by reducing DNA nicking. bioRxiv 5: 574020. doi:https://doi.org/10.1101/574020

14. Niu D, Wei H-J, Lin L, et al (2017) Inactivation of porcine endogenous retrovirus in pigs using CRISPR-Cas9. Science 357: 1303–1307. doi:https://doi.org/10.1126/science.aan4187

15. Łopata K, Wojdas E, Nowak R, et al (2018) Porcine Endogenous Retrovirus (PERV)—Molecular Structure and Replication Strategy in the Context of Retroviral Infection Risk of Human Cells. Front Microbiol 9: 432. doi:https://doi.org/10.3389/fmicb.2018.00730

16. Wright DWM (2018) Cloning animals for tourism in the year 2070. Futures 95: 58–75. doi:https://doi.org/10.1016/j.futures.2017.10.002

17. Folch J, Cocero MJ, Chesné P, et al (2009) First birth of an animal from an extinct subspecies (*Capra pyrenaica pyrenaica*) by cloning. Theriogenology 71: 1026–1034. doi:https://doi.org/10.1016/j.theriogenology.2008.11.005

18. Crichton M (1991) Dinopark. Droemer Knaur Verlag, München

19. Griffin DK, Larkin DM, O'Connor RE (2019) Time lapse: A glimpse into prehistoric genomics. Eur J Med Genet. doi:https://doi.org/10.1016/j.ejmg.2019.03.004

20. Ro D, Paradise E, Ouellet M, et al (2006) Production of the antimalarial drug precursor artemisinic acid in engineered yeast. Nature 440: 940–943. doi:https://doi.org/10.1038/nature04640

21. Hommel M (2008) The future of artemisinins: natural, synthetic or recombinant? J Biol 7: 38. doi:https://doi.org/10.1186/jbiol101

22. Peplow M (2016) Synthetic biology's first malaria drug meets market resistance. Nature 530: 389–390. doi:https://doi.org/10.1038/530390a

23. Westfall PJ, Pitera DJ, Lenihan JR, et al (2012) Production of amorphadiene in yeast, and its conversion to dihydroartemisinic acid, precursor to the antimalarial agent artemisinin. Proc Natl Acad Sci USA 109: E111–8. doi:https://doi.org/10.1073/pnas.1110740109

24. Paddon CJ, Westfall PJ, Pitera DJ, et al (2013) High-level semi-synthetic production of the potent antimalarial artemisinin. Nature 496: 528–532. doi:https://doi.org/10.1038/nature12051

25. Hutchison CA, Chuang R-Y, Noskov VN, et al (2016) Design and synthesis of a minimal bacterial genome. Science 351: aad6253. doi:https://doi.org/10.1126/science.aad6253

26. Elowitz MB, Leibler S (2000) A synthetic oscillatory network of transcriptional regulators. Nature 403: 335–338. doi:https://doi.org/10.1038/35002125

27. Pedersen M, Phillips A (2009) Towards programming languages for genetic engineering of living cells. J R Soc, Interface 6: S437–S450. doi:https://doi.org/10.1098/rsif.2008.0516.focus

28. Sturtevant AH (1923) Inheritence of direction of coiling in *Limnaea*. Science 58: 269–270. doi:https://doi.org/10.1126/science.58.1501.269

29. Tumpey TM (2005) Characterization of the Reconstructed 1918 Spanish Influenza Pandemic Virus. 310: 77–80. doi:https://doi.org/10.1126/science.1119392

Further Reading

Buddingh BC, van Hest JCM (2017) Artificial Cells: Synthetic Compartments with Life-like Functionality and Adaptivity. Acc Chem Res 50: 769–777. doi:https://doi.org/10.1021/acs.accounts.6b00512

Church GM (2012) Regenesis. Basic Books, New York/USA.

Kuldell N (2015) Biobuilder. O'Reilly Media, Sebastopol, California/USA.

Sleator RD (2016) Synthetic biology: From mainstream to counterculture. Arch Microbiol 198: 711–713. doi:https://doi.org/10.1007/s00203-016-1257-x

7
Genes and Society

I experienced a big surprise in February 2010 when I came across a kiosk of a special kind in the German city of Weimar (Fig. 7.1). Not only did it spell itself with a "Y", no, it also advertised itself with a motto modified from the pharmaceutical company *Bayer*: Instead of *Science for a better Life*, the kiosk read *DNA for a better Life*. Aha?

Those who could foresee that they would have problems with a paternity test or a dragnet using genetic fingerprints could acquire a new identity there for a corresponding fee—by acquiring DNA from a "clean" donor. The donor DNA came from law-abiding people without entries in the criminal record or the Interpol register. On closer inspection, the project turned out to be an installation by the action artists Oleg Mavromatti from Moscow and Bionihil from Weimar. The DNA kiosk was intended as an invitation to reflect on the link between science and art on the one hand and science and society on the other. The application of genetic analysis and genetic engineering as citizen science is now widely established, whether in the context of the *quantified-self-movement,* that is self-measurement, or as an attempt to set a counterpoint to the commercial use of genetic engineering. We should also be interested in the

© Springer-Verlag GmbH Germany, part of Springer Nature 2022
R. Wünschiers, *Genes, Genomes and Society*,
https://doi.org/10.1007/978-3-662-64081-4_7

Fig. 7.1 A kiosk in Weimar where you can get DNA for a better life?

way in which we can protect and perhaps even archive the totality of all genetic information on Mother Earth.

7.1   Citizen Science

For a long time, science was something for people who had acquired special knowledge and applied it to answering a wide variety of questions. Non-scientists could at best participate passively, for example by providing computing power from their private computers for projects such as SETI@home (*search for extra-terrestrial intelligence at* home) or the search for a vaccine against Ebola viruses. Over 72,000 years of computing power have already been donated to the latter project in this way since December 2014. But many citizens also contribute to science indirectly and

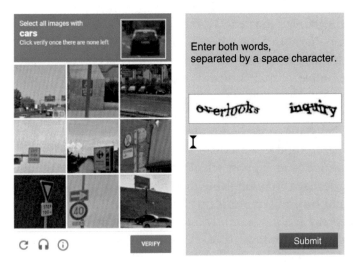

Fig. 7.2 These CAPTCHA queries, designed to distinguish humans from automata, often involve users in the solution of scientific problems, in this case, the optimization of image recognition software, without being asked

often unknowingly. For example, mobile phones help to analyse and predict traffic flows. And the famous CAPTCHA technology helps companies like *Google* to improve their software for text or image recognition (Fig. 7.2). This, in turn, should help to ensure that the digital archiving of written documents or recognition software for autonomous driving is constantly optimised (Sect. 8.2). **CAPTCHA** stands for *completely automated public Turing test to tell computers and humans part,* that is an automated test to distinguish humans from computers and thus prevent so-called *bots* (short for robots) from automatically filling out forms on the Internet.

However, it is also possible to become privately active in science on one's own initiative. This involvement is known internationally as **citizen science**. For example, the

Argentinian amateur astronomer Victor Buso observed a supernova in September 2016 and became a co-author in the renowned science magazine *Nature* [1]. Currently, for example, projects aimed at recording **biodiversity** are flourishing. Numerous nature conservation associations and societies have launched calls to report animals, insects or plants to a central office. Here the data are checked, summarised and made available to universities and research institutions to support scientific projects. Hobby ornithologists have been supplying their observations to associations and **bird observatories** for decades. The *German Weather Service* is also constantly looking for volunteers to observe the weather and also provides the measuring equipment for this purpose.

Similarly, owners of over 3000 **dogs** contributed to the identification of the genetic basis for bright blue eyes in Siberian Huskies [2]. In my own research, I also rely on the willingness of owners of sheep poodles who voluntarily send me a blood sample from their dogs via their veterinarian. In my research group, we use the samples to develop genetic tests for certain diseases.

A popular activity for many people is to trace their own **ancestry.** For this purpose, there are many free portals on the Internet. The first and easiest step is to enter your own relatives to get a descriptive family tree. However, the data can also be expanded to include genetic components. Computer programs search in the background for similarities (for example, in birth and death dates of ancestors) in different profiles and suggest new potential relatives in this way. In 2018, a group of researchers led by the US computer scientist Yaniv Erlich of Columbia University created the world's largest family tree to date using one-tenth of the data from the *geni.com* platform and made the data available anonymously on *familinx.org* [3]. It spans all continents and a period of 500 years, or 11 generations. All the data was fed into the platform by amateur genealogists in the years before. From this, interesting facts could be gleaned. On the basis of the ages of the persons, for example, it was possible to reconstruct the two world wars, but also the decline in infant mortality in the twentieth century. These results were of course to be expected and served more as a quality control of the data set. It should be noted that 85% of the data came from Europe and North America. The dataset brought new insights into the inheritance of **life expectancy**. The scientists were able to conclude that the genetic contribution to life expectancy is about 16%, which is considerably lower than the 25% previously assumed. Thus, according to this study, the contribution of the environment or the way of life is more predominant than previously assumed. Interesting data on migration were also found: According to this study, in the past 300 years it was mainly men who moved long distances in North America and Europe and whose country of birth was different from that of their children. Similarly, the distance between the birthplaces of spouses has increased from around 10 kilometres at the beginning of the nineteenth century to 100 km now.

Due to the increasing interest of citizens in genetic diagnostics, both in terms of genealogical research and in medical terms (Sect. 7.3), public databases are filling up more and more with genetic data. In 2017 alone, around seven million DNA analysis kits were sold to private households in the USA [4]. The cost per kit starts at around €60. After swabbing your cheek with a cotton swab and returning the sample, you will have your own genetic profile in about 4 weeks. The three largest US **genetic analysis providers**, *23andMe, Family Tree DNA* and *Ancestry,* now have more than 15 million users. The data does not (yet) include the complete genome, but around 720,000 selected markers that are associated with certain characteristics or diseases and are used for diagnostic purposes (Sect. 4.2). However, relationships can also be derived from these data, which is why many users additionally make their genetic data available to the Florida-based online service *GEDmatch*. Over 600,000 interested people have already done so. Although the user accounts are not public, **law enforcement agencies**, for example, can use them—users agree to this in the terms of use [5]. Among them was a relative of the so-called Golden State Killer Joseph DeAngelo from California. He was blamed for 13 murders and more than 50 rapes between 1974 and 1986, but was never caught. In early 2018, detectives uploaded the DNA profile of a 37-year-old DNA sample from one of the crime scenes to *GEDmatch* [4]. This identified nearly 20 potential relatives who were also registered in the database, including DeAngelo's third cousins (Fig. 4.12). Research in the vicinity of these possible relatives eventually led the trail to DeAngelo, who was subsequently convicted on the basis of DNA testing. A major contribution to the identification of the culprit was made by the US biologist and genealogist Barbara Rae-Venter, who was named one of the ten most influential scientists of 2018 by the science magazine *Nature* (as was He Jiankui, Sect. 3.2) and who was the first wife of the world-famous genetic technologist Craig Venter (see also Chap. 4 and Sect. 7.2).

The wealth of data in the databases of ancestry and gene analysis providers is already so large that with a 60% probability at least one eighth-degree relative (Fig. 4.12) can be found for all North Americans or Europeans via a DNA profile. With the addition of further information such as place of residence, gender and age, a person can currently be assigned almost directly, at least in the USA (Fig. 7.3).

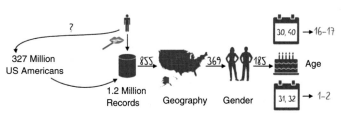

Fig. 7.3 Via DNA profile matching with public data records, 855 relatives can be assigned to a person. With additional information such as place of residence, gender and age, a person can currently be almost uniquely identified in the USA. The genetic data is not the forensic genetic fingerprint, which is not publicly available

Genetic diagnostics as citizen science, with or without the participation of scientists, is an interesting and certainly largely harmless development. But what about genetic engineering, the active intervention in the genetic material of microorganisms, plants, animals or humans? In the field of genetic engineering, a community of interested people who call themselves *do-it-yourself biologists* or **biohackers** has been developing for some years now. In so-called *live hack spaces* (laboratories), **DIY biologists** try to tackle more or less everyday questions. This also includes genetic experiments. In the USA, laboratories with full equipment can be rented cheaply from some institutions or interest groups. Others work in a hobby cellar or in a garage that has been converted into a private laboratory, which is why people like to talk about **garage laboratories**. This also refers to the time in the 1970s when pioneers like Bill Gates with Steve Ballmer founded Microsoft and Steve Jobs with Steve Wozniak founded Apple and turned the computer into a personal computer.

The nimbus of innovative garage science always has something hidden about it. And this is where the problem lies when it comes to genetic engineering. On the one hand, it is to be welcomed that people interested in genetic engineering can be found. The difficulty is that downloading instructions on how to genetically modify an organism cannot be compared to a baking recipe. A failed meal may scare guests, but a failed genetic engineering experiment and pouring the leftovers down the sink can have unforeseen consequences. At the very least, it is about assessing the risk and proceeding responsibly accordingly. In 2017, an **experimental kit** from the US company *ODIN* caused a stir in Germany (Fig. 7.4).

The box cost US$159 and contains all the components needed to genetically modify a bacterium using the CRISPR/Cas system (Sect. 5.1). During 10 h of experimentation, which are spread over 2 days because of the bacteria cultivation, a bacterium (the intestinal bacterium

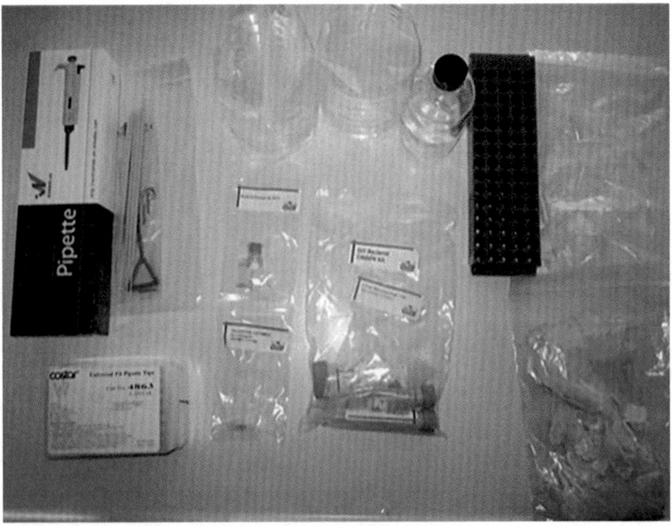

Fig. 7.4 The experimental kit from the US company *ODIN* in my laboratory. That is all it takes to genetically modify cells with the CRISPR/Cas system. (Photo: Robert Leidenfrost)

Escherichia coli, which has been very well studied for decades) is made resistant to the antibiotic streptomycin. The bacterium used is a safety strain that cannot colonize the human gut, nor is it viable outside the lab. The experiment itself is based on research results from 2013 [6]. In the US the experiment is not regulated and there are videos on *YouTube,* for example, where the experiment is carried out in the kitchen, so the kitchen becomes a ***biohack-space***. In Germany, this is not allowed, but requires a level 1 genetic engineering safety lab and knowledgeable scientists. Otherwise, a fine of up to €50,000 can be imposed, or even a **prison sentence** of 3 years if a GMO is released. The fact that this is a good thing became apparent at the end of 2016 when the *Bavarian State Office for the Environment and Consumer Protection* discovered that pathogenic representatives were growing alongside the expected *Escherichia coli* bacteria [7]. Apparently, the bacteria supplied were **contaminated**. As a result, the import of the experimental kit was regulated. The *European Centre for Disease Prevention and Control* (ECDC) conducted a risk assessment in May 2017 and concluded:

> [...] the potential contribution of the contaminated kit to the increasing burden of antimicrobial resistance in the EU [...] is low and the associated risk to public health is considered to be very low.

So, caution is advised, but there is no hazard. But the incident also revealed something else: The *Bavarian State Office* denied German DIY biologists access to the detailed results. They then turned to the renowned *European Molecular Biology Laboratory* (EMBL) in Heidelberg, where they were able to describe the contamination by DNA sequencing in safety laboratories and with expert scientists. It is this **open dialogue** between researchers and DIY biologists that is a welcome breath of fresh air in the genetic engineering debate in particular and the openness of science in general.

However, despite all openness in dealing with genetic engineering, it must be clearly defined where the limits are. For example, 28-year-old HIV-positive Tristan Roberts from the USA gave himself gene therapy in a *Facebook* live stream on October 18, 2017 (Fig. 7.5). In front of a running camera—the treatment was streamed live on *Facebook*—he injected himself with an unapproved therapeutic agent from the Singapore company *Ascendance Biomedical,* which incorporates a DNA fragment into his body cells, whereupon these form an antibody as the active ingredient.

In 2016, studies by scientists at the *US National Institutes of Health* (NIH) showed that this antibody, called N6, blocks a potential entry site for the HI virus, the so-called **CD4 antigen** (a glycoprotein consisting of a sugar (glyco) and a protein component), thus making it invisible to the virus [8]. It should be noted at this point that the CD4 antigen, together with the CCR5 receptor on the white blood cells (T cells) of the immune system, serves as an entry site for the HI virus (Sect. 5.1 and Fig. 5.3) [9]. Since then, research has been conducted in clinical trials to

Fig. 7.5 Public livestream on *Facebook* on October 18, 2017: Tristan Roberts injects himself with an unapproved gene therapy drug for AIDS

develop a suitable drug. DIY researchers did not want to wait for its completion. After the injection, the injection site on Tristan Robert's abdomen became inflamed—the therapy itself did not work. Since May 2018, Roberts has been treated conventionally again with *highly active anti-retroviral therapy* (HAART). This combination therapy, introduced back in 1996, uses up to four antiretroviral agents. Roberts has not yet given up hope of finding a biohack to fight AIDS. On the Internet, he directly addresses biohackers with experience in working with the CRISPR/Cas system—they may contact him.

There are numerous similar cases and the motivation ranges from a (possibly understandable) **DIY AIDS therapy** to the hope of eternal youth through the introduction of gene constructs that cause cells to produce corresponding hormones. In 2018, the US *Food and Drug Agency* (FDA) issued a very official warning against such private genetic manipulations [10].

This does not stop biohackers and transhumanists from researching in private and making big plans. **Transhumanists** aim to expand the possibilities of the human body and mind with technology. Civilian researchers exchange ideas on online platforms. One well-known platform, the **DIYhplus Wiki**, brings together 62 groups spread around the globe, including from Berlin and Munich. It is a source of *open-source* information from biohackers for biohackers. The wiki also contains instructions for DNA synthesis and gene editing, including method optimisation. Based on the motto *biology is technology*, ethical questions are not even asked. On the contrary. In contrast to the precautionary principle (Sect. 3.7), the operators of the wiki call for **proactive action**:

> The freedom of people to be technologically innovative is extremely valuable or even critical to humanity. This implies

a number of responsibilities for those who consider whether and how to restrict new technologies.

Accordingly, it is not the users of genetic engineering who should justify themselves, but the regulatory authorities.

One goal of the transhumanists is to increase **memory**. Would not it be great not to forget anything anymore? But how does our memory actually work in the brain? Where in our 85 billion nerve cells (depending on the source, the number varies—there are many) is the memory located? The German psychologist Hermann Ebinghaus founded experimental memory research with his work on learning and forgetting at the end of the nineteenth century [11]. A seemingly self-evident but previously unquantified result of his experiments was the importance of repetition in learning. In 1949, the Canadian psychologist Donald Hebb clarified the neural basis for the phenomenon described by Ebinghaus, the so-called Hebb's rule:

> If an axon of cell A excites nearby cell B and repeatedly and persistently contributes to the generation of action potentials in cell B, this results in growth processes or metabolic changes in one or both cells that cause the efficiency of cell A to become greater with respect to the generation of an action potential in B [12].

Learning thus ultimately causes a molecular change. An important role is played by a protein that absorbs neuronal stimuli in the membrane of neurons, the N-methyl-D-aspartate (NMDA) receptor. One hypothesis pursued by scientists was that the more NMDA receptors there are in a cell, the better the neuron or better, its owner should be able to learn. That this is indeed the case was proven in mice in 1999 [13]. Surprisingly, the genetically modified mice with more neuronal NMDA receptors not only learn better, but they also have a better memory. However, this also comes at a price: once learned, the mice no longer forget. As a result, after a certain time the memory is full and nothing new can be learned. It is a good thing that the transhumanists have not yet carried out this genetic intervention on themselves. So, we see that although memory can be expanded, its dynamism is lost. Research in this direction is not standing still. The US military research institution DARPA (*Defense Advanced Research Projects Agency*) has a great interest—not in augmentation, but—in restoring damaged brains and their memory in soldiers. Corresponding research projects are currently being funded by the US Department of Defence.

The US biohacker Gabriel Licina has the idea of accelerating the growth of trees by amplifying the activity of two genes, *PXY and CLE41*. This is a project he calls aggressive **reforestation** or engineered tree systems. In birch trees, it has been shown that the activity of the genes causes them to grow twice as fast. Licina's goal is to make the technology available to everyone. What drives him? He says:

> There's the stuff that pays the bills [...] and then there's the stuff that you really care about.

If you want to get an impression of the ideas and developments of biohackers and transhumanists, just take six hours and watch the recording of the last *BioHack-the-Planet* conference 2018 from Oakland/USA on *YouTube* [14].

7.2 Commercializing Genetic Information

As should have become clear from the previous sections, the money is rolling in at full speed in the field of marketing genetic information and genetic methods [15]. It probably occurred to few that when the first draft of the DNA sequence of the human genome was announced in 2000, two versions were actually presented. The representatives of these two versions were standing in the White House on 26 June next to the then US President Bill Clinton: on the one side the geneticist and entrepreneur (*Celera* company) Craig Venter, on the other the geneticist Francis Collins as representative of the international, publicly funded *Human Genome Organisation* (HUGO). HUGO was the largest research project worldwide up to that time. It was a publicly funded alliance of more than 1000 researchers from 40 countries with the declared goal of deciphering the 3.2 billion nucleotides of the human genome by 2015 and making them publicly available. Venter left HUGO as early as 1992, founded a research institute and pursued his own sequencing activities in parallel with the public project. Germany joined HUGO in 1995 and contributed almost 60 million bases to the human genome via the participating research institutes in Berlin, Braunschweig, Heidelberg and Jena.

While the DNA sequences from HUGO are freely available, customers had to pay money for the *Celera* dataset. Later, Craig Venter offered numerous other genomic datasets for sale, for example completely sequenced **ecosystems**. Of great importance, however, was Clinton's announcement in 2000 that the human genome was not patentable: it caused biotechnology stocks to crash.

The issue of patenting is a controversial one. Can DNA sequences be patented? A major controversy arose when the biotechnology company *Myriad Genetics* was granted European patent number EP699754 on the DNA sequence of the *BRCA1*-gene and a genetic test in 2001. Approximately 10–15% of all **breast cancers** are caused by mutations in the *BRCA1* and *BRCA2* genes. *Myriad Genetics* had a monopoly on the diagnosis of the gene variants through patenting and at times charged over US$4000 for the service. In 2004, the **European Patent Office** ruled that the company's original application did not relate to a novelty, but only to a DNA sequence, and revoked the patent. *Myriad Genetics* appealed, however, and in 2008 was granted a modified patent that covers testing for specific mutations, but not the gene itself. In 2013, the US patent was also rejected on the grounds that isolating sequences is not enough to patent them. Against the background that **Big Data** has also reached the life sciences and that DNA sequence data from patients are being generated on an unimaginable scale (for example, in the English BioBank project), it becomes clear, however, that knowledge is power (Sect. 8.2) [16]. If association studies reveal the connection between diseases and genes, variants of these genes (alleles) or epigenetic changes (Sect. 8.1), then this information can be used to develop lucrative genetic tests and to sell them as a service. The patients and not the patent must always be in the foreground. It is therefore essential for industry to agree on clear rules with policy-makers who serve the common good, so that knowledge acquired in industry also benefits the general public.

However, the fact that even this is not easy is shown by the example of **Golden Rice**, a rice variety that contains increased amounts of provitamin A (beta-carotene) through genetic engineering. The development has been started in 1992 by the German biologist Ingo Potrykus and the cell biologist Peter Beyer and was published in 2000 [17]. The stated aim is to combat the vitamin A deficiency prevalent in many developing and emerging countries. Golden Rice, named after the golden colour of the rice seeds due to the high vitamin A content, is regarded by some as a showcase project, but by others as a kind of Trojan horse of plant genetic engineering. Although this rice is not yet marketed, mainly for political and ideological reasons, the US Patent Office awarded it the *Patents for Humanity Award* in 2015. This honours the release of patented technologies for global humanitarian applications. This award points to a special feature of the Golden Rice Project: It is intended to serve the common good and the seeds are to be given away without royalties. It sounds like a dream, but in the dispute about genetic engineering in general and the form of development aid to be provided in particular, it has become a nightmare for Potrykus and Beyer. Nevertheless, research is progressing and field trials are taking place in the USA, Vietnam and the Philippines, among other countries, and are also financed by the private *Bill and Melinda Gates Foundation*. In Australia and New Zealand, the import of Golden Rice has been approved for some time. Similarly, Canada and the US have approved imports since early 2018, and more countries are looking to follow suit. In addition, Golden Rice is being further developed. In 2018, for example, a Chinese research team presented aSTARice, a further development that forms other antioxidants in addition to provitamin A (Fig. 7.6) [18].

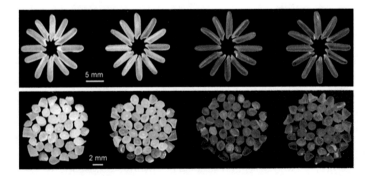

Fig. 7.6 Rice grains of different varieties. From left to right: wild rice, golden rice, canthaxanthin rice and astaxanthin rice (aS-TARice). The top row shows the whole grains and the bottom row shows the broken grains. (From Zhu et al. [18])

In general, the path from the idea of generating a genetically modified organism to an approved commercial product is a very long and costly one. For this reason, it is not surprising that the seed business with GMO plants, for example, is dominated by a few **large corporations** such as *Bayer* (which has swallowed *Monsanto*), *Syngenta, DuPont Pioneer, BASF* or *Dow*. It is difficult to afford an **approval procedure** with the necessary and verifiable preliminary tests and studies to be provided. For example, it must be demonstrated in the application that the GMO has no adverse effects on humans, animals or the environment. In the case of food or feed, analyses must show that the GMO food does not differ significantly from conventional comparative products and does not contain any allergens—a commercial kiwi would not pass this test. In addition, **in-market monitoring** procedures must be submitted to identify the organism. The application is then forwarded to the *European Food Safety Authority* (EFSA) for consideration. The EFSA may require the applicant organisation to carry out further investigations, which are often costly and lengthy. The national authorities of the member states are involved in the procedure and can in turn request additional data. The *EU Reference Laboratory* validates the methods proposed by the applicant to detect and identify the GMO in question. Finally, EFSA forwards an opinion to the *EU Commission* and the Member States and makes it available to the public. The Commission then submits a proposal for a decision to the Member States. A qualified majority is required for adoption. This is the case if 55% of the member states (currently 15 out of 28) agree and 65% of the EU population is represented at the same time. At best, the procedure takes 9 months, but usually several years. This requires a capital buffer on the part of the applicants in order to survive the approval process. It is understandable that corporations recoup this investment from the farms in the form of licence fees. The situation is similar for GMO-free **conventional seeds**. Here, too, long breeding work goes into the seed and an approval procedure at the *Federal Plant Variety Office* and, if necessary, at the European *Community Plant Variety Office* (CPVO). Here, too, farmers have to pay licensing and **cultivation fees**, especially for hybrids.

As already described, in hybrid breeding, suitable, separately bred inbred lines are crossed with each other once (Sect. 3.2). The offspring (first generation) of such a cross often have agronomically valuable properties compared to the parent generation, such as stronger growth or larger fruits (heterosis effect). However, further breeding of the **hybrid variety** is not economical, since sowing the seeds of the first generation in the second generation again produces the parental characteristics—the profitable **heterosis effect** is lost. Instead, the farm again has to buy new seed. To counteract the growing privatisation in the seed sector, an alliance of breeders and lawyers has developed an **open-source licensing model** for seeds. This is based on comparable models in the software sector and only allows use on condition that users/breeders undertake to make further developments available under the same licence model and not to patent them. Currently, seven free varieties are available in Germany: three tomato varieties, three wheat varieties and one maize variety. Whether the concept will catch on remains to be seen, as the development costs for new varieties or breeds must at least be borne.

In the context of discussions on the exploitation of farmers with patented, genetically modified genetic material, a clear distinction must be made between the technology on the one hand and the economic exploitation on the other.

Science must assess and minimise the risk of a technology. It is also a question of **business ethics** to assess the consequences of business models for people and the environment and to identify viable options. This is also the result of the so-called **Monsanto Tribunal**, which took place in The Hague in 2016/17. Five judges set out in a legal opinion how *Monsanto's* (now *Bayer*) practices violate human rights and lead to ecocide, among other things. The Tribunal was not an officially recognized court, as there is currently no legal instrument to prosecute companies and its directors as responsible for crimes against human health or against the integrity of the environment.

With regard to patenting, however, it is also important to know that there is a **territoriality principle**. This means that patents are only valid in the country for which they were granted. Countries can therefore defend themselves against patents by not allowing them in the first place. Thus, in developing countries, most varieties are not patented. This, of course, usually leads to companies trying not to serve the relevant market—if they can afford to do so. On the other hand, dependencies prevail. India, for example, only confirmed a patent on BT cotton in January 2019 to boost trade [19]. As a member of the **World Trade Organization** (WTO), India has to comply with the *Trade-Related Aspects of Intellectual Property Rights (*TRIPS) Agreement [20]. In this way, of course, pressure is put on states. Perhaps *the* big question in relation to our food base is how we regulate or deregulate or re-regulate international markets in the future.

Another aspect is the marketing of my very own genetic information. Large corporations like *Google, Facebook, Apple* or *Amazon,* but also small ones like *FitBit, Garmin* or *Suunto,* are continuously collecting data of their users. They are already earning a lot of money with it. What would it be like if internet, communication, shopping, exercise and **sports behaviour** were linked to genetic information (Sect. 8.2)? The genetic data could then be provided by platforms such as *23andMe* or *Ancestry* (Sects. 4.3 and 7.1*)*, the shopping behaviour by *Amazon* and *eBay*, the fitness data by *Garmin, Strava* or *Apple*. Personalised genetic data are generally not freely available, but we have seen in Sect. 7.1 (see also Fig. 7.3) how comparatively easy it is to assign data to a person. This is already common practice in the field of medicine. In 1998, the Icelandic government commissioned the company *deCODE Genetics* from Reykjavík in Iceland to carry out a comprehensive collection and storage of all health data of the population [21]. The genome data of more than 2600 Icelanders belong to the company and are sold exclusively. The largest sequencing project to date is likely to be similar: The *National Health and Medicine Big Data Nanjing Center* in China's Jiangsu province announced in October 2017 that it would sequence the genomes of one million Chinese [22]. But most of these projects are publicly funded. A number of scientists have therefore issued a joint letter calling for genomic data, from bacteria to humans, to be made publicly available on appropriate platforms [23]. So far, this is only partially the case, and often the raw data are poorly described, so that they can hardly be used for further studies (Sect. 8.2).

7.3   My Genes and Me

I am a part of society and my genes are a part of me and—I am more than my genes. We should keep that in mind before we start this section. Almost all of us carry a smartphone with us every day that is equipped with all kinds of sensors. In addition, there are smartwatches and fitness trackers of all kinds, which are no longer missing from any shopping brochure. They serve a more or less self-reflective **quantified-self-movement**, that is a movement of people that like to measure all aspects of themself. It is often forgotten that correlation does not imply causality. That is, not every measurement makes sense. Increasingly, the quantified-self-movement also makes use of the possibilities of genetic diagnostics or, as we have seen in Sect. 7.1, even gene therapy.

In the field of personalized medicine, personal data can even be very helpful and will certainly gain in importance in the future. It is possible that we will then take them with us to the doctor's visit along with our health card. The US biologist Leroy Hood not only coined the term systems biology as the comprehensive view of a biological system, but also that of **P4 medicine** [24]. This form of medical prevention and therapy is supposed to be *predictive, personalized, preventive* and *participatory.* Patients and their data therefore play a major role. But where do we currently stand with quantified-self?

Lifelogging and quantified-self-tracking are the magic words of the generation gene editing. The immediate goal is the optimization of being human—by optimizing one's own life circumstances, that is a kind of do-it-yourself-enhancement. However, this is not achieved by means of a well-thought-out organization of daily appointments and a balanced diet, but by recording as much data as possible about one's own activity with mobile devices, the activity trackers. Worn as a bracelet, they use acceleration sensors to record at least body movement (usually also during sleep), but sometimes also geographical position and altitude, temperature, air pressure and heart rate, and transmit the data wirelessly to a smartphone. They can be supplemented with consumed coffees, vegetable soups or other. Then an app does the data analysis. Some wristbands allow information to be fed back to the wearer, for example via vibration or visual signals. In this way, users can be informed about a completed distance or be alerted to indications from the smartphone program. This program constantly evaluates the data; not autonomously, but usually via the Internet with a central server. Once the app has analysed the users and their behaviour for a few days—one might say: got to know them—then evaluations and rules for action are displayed, for example of this kind: *You spend too much time reading in bed; go for a longer walk*, etc. Of course, you are relying on the app's programmers and their definition of too much and too little of something. Combining personal medical data with data from the app and one's own sense of well-being quickly results in an amateurish medical history with cybermedical treatment à la *Google*. Medication is given at the click of a mouse from the Internet mail order system. Doctors are then actually only needed for sick notes. Despite all the benefits, the risk of incorrect treatment seems to outweigh the benefits.

As a social being, people usually like to share and motivate themselves. A typical application of lifelogging is therefore the posting of sporting activities and delicious meals on social networks on the Internet. And although the mass accessibility of this personal enhancement by means of self-promotion through media technology is novel, it is not new. The publicly celebrated plastic surgery of the US actress Fanny Brice was already sensational in 1923. What is new is that masses of data can be recorded in parallel and correlated with each other. But lifelogging is by no means the limit of what is already possible. Increasingly, **fitness scores** are being supplemented by **genomic scores** [25]. Even one's own genetic material has become available to everyone thanks to modern and inexpensive sequencing methods (Chap. 4). The American company *23andMe,* founded in 2006, offers anyone a genetic genotype analysis based on a saliva sample for currently €169 [26]. On the basis of this data, customers receive ...

- ... an **ancestry report** including his relationship to Neanderthal man.
- ... a **health report** on health risks such as for breast cancer and Parkinson's disease.
- ... a **wellness report**, including a genetically predicted optimal weight (the so-called genetic weight) or the presence of lactose intolerance.
- ... a **condition report** on thirty different phenotypic traits such as the condition of the earwax, the eye colour or the tendency to get up early or late.
- ... a **hereditary disease carrier status** regarding 43 diseases such as sickle cell anaemia.

The really explosive thing about private genome analysis is that customers are left alone with the data they retrieve on their personal websites. This violates the German **Genetic Diagnostics Act** (*Gendiagnostikgesetz*), according to which only doctors with proven expertise are allowed to open and explain the findings to patients. And there is a reason for this: For example, genetic evidence of Huntington's disease (a hereditary disease of the brain) has a high predictive power, whereas a detected mutation in the *BRCA-1* gene associated with breast cancer has only a low predictive power with regard to the onset of the disease. This is because often not only one, but often several genetic components or even epigenetic and environmental factors (Sect. 8.1) play a role.

The next step is to sequence not only us, but also our fellow inhabitants. I do not mean the human ones, but the microorganisms on our skin and especially in our digestive tract. We have more bacterial cells in our gut than we have cells of our own. Why should we analyse the so-called **gut flora**, more correct would be actually intestinal fauna? As is well known, the way to the heart goes through the stomach. But it can also hit the stomach or, to paint a more positive picture, make butterflies flutter in the stomach.

A bundle of nerve fibres between the sternum and the navel, known as the *solar plexus*, is responsible for the latter. It is directly connected to the oldest and deepest part of the brain, the so-called brain stem or reptile brain, and regulates vital functions such as breathing, the regulation of the heartbeat and intestinal activity. At least as far-reaching, however, is the indirect connection between the stomach, or more precisely the intestine, and the brain via the so-called microbiome-gut-brain axis. The **microbiome** is the totality of all microorganisms, in this case in the intestine [27]. This is neither stable in terms of the up to 1000 microorganisms involved, nor in terms of their quantitative relationship to each other [28]. That a connection exists between the bacteria in the intestine and the brain was demonstrated by the US microbiologists Linda Hegstrand and Roberta Jean Hine as early as 1986 with sterile reared rats [29]. Today we know that neuroactive substances such as the neurotransmitters histamine and GABA (gamma-aminobutyric acid) are produced in the intestinal flora. These can affect not only the response to stress, but also cognitive abilities and probably even mental health. Diseases of the central nervous system such as multiple sclerosis, Alzheimer's disease, schizophrenia, autism and depression are suspected to be influenced by the gut flora [30]. The complexity of the relationships can be appreciated when one considers that the 2018 version 4.0 of the human metabolome database (HMDB) already lists 114,100 substances [31].

Companies such as *for me do* have long been offering genetic analyses in order to optimally nourish oneself and thus support one's own **fitness** and training. For this purpose, numerous genes are examined for the variants present (Fig. 2.3). These genes were selected because they code for proteins and enzymes that are related to specific metabolic diseases. After this *MetaCheck-fitness analysis,* clients are classified into one of four different **meta-types** (alpha, beta, gamma, and delta). Further subdivisions are made according to whether one is more of an endurance athlete or a power athlete. In fact, for two genes in particular, there is a proven connection between certain gene variants and athletic performance—at least in competitive athletes [32]. For example, there are two variants of the angiotensin-converting enzyme ACE that are correlated with performance. The enzyme plays a major role in maintaining blood pressure and regulating water and electrolyte balance. The other is the *ACTN3* gene, the product of which is involved in building muscle (Fig. 2.3). But what will be the effects if our diet no longer depends on **intuition** but on genetic analysis results?

A current example, again from the nutrition sector, should illustrate the importance of the topic. We live in a time in which awareness of healthy food is widespread. However, this not only leads to the preferred consumption of wholesome foods, but also to the supposed upgrading of foods with additives. These **functional foods***,* propagated by the food industry, are supposed to protect against deficiency symptoms and give consumers a healthy lifestyle. In some countries, the artificial addition of vitamins and other nutrients to food is even prescribed by law. In Europe, for example, vitamins must be added to baby food, and in the USA, as well as dozens of other non-European countries, folic acid must be added to flour [33]. In principle, **folic**

acid is an essential component of food and is recommended to pregnant women in the form of tablets as a food supplement. Folic acid deficiency has been linked to a number of diseases, including various types of cancer and Alzheimer's disease. People who carry a rare variant (allele) of an important enzyme, methylenetetrahydrofolate reductase (MTHFR), of the folic acid metabolism are particularly affected (Fig. 7.7). In this allele, called *677 T-MTHFR*, a cytosine is exchanged for a thymine at position 677 of the gene sequence. This exchange affects the second position of the alanine-coding triplet GCT, which now codes for valine as a result of this base exchange.

Carriers of this gene variant (about 0.2% of the US population) are significantly more likely to suffer from diseases otherwise caused by folic acid deficiency. Therefore, the disease of these people is prevented by administering increased

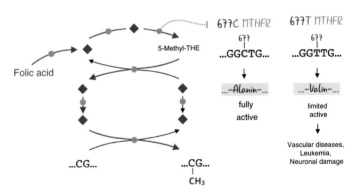

Fig. 7.7 Folic acid is one of several sources of DNA methylation. A kind of reaction equilibrium is at work: The more folic acid is present in the diet, the more DNA can be methylated. Involved in the reaction chain is an enzyme called methylenetetrahydrofolate reductase (MTHFR). A SNP (Sect. 4.1) at position 677 in the coding gene can lead to an altered amino acid in the enzyme and thus to restricted activity and disease. Green circles: enzymes; blue diamonds: metabolic intermediates

doses of folic acid with the diet. However, a study from 2005 was able to prove that a constantly increased intake of folic acid leads to a preferential selection of the defective gene variant in the population, since carriers of the defective allele no longer show symptoms [34]. This means that single individuals who are carriers of the defective gene variant are helped by increased folic acid supplements. For the population as a whole, however, increased folic acid administration has a negative effect—an effect that will not be observed in the short term, but only over several generations.

We are thus in a field of tension between individual and public healthcare, based on the statistical analysis of an unmanageable amount of data by partly opaque computational processes (algorithms, Sect. 8.2).

7.4   Gene Banks

Since its entry into force in 1993, the *United Nations Convention on Biological Diversity* (in short: **Biodiversity Convention**), which has since been signed by 196 countries, has provided a binding legal framework for the protection of biological diversity at international level (Fig. 7.8). It covers the protection of genetic, species and ecosystem diversity. The protocol also points to risks associated with the deliberate and unintentional **release** of genetically modified organisms, which is why each partner

> Establish or maintain means to regulate, manage or control the risks associated with the use and release of living modified organisms resulting from biotechnology which are likely to have adverse environmental impacts that could affect the conservation and sustainable use of biological diversity, taking also into account the risks to human health.

Fig. 7.8 Biodiversity is not only beautiful, but also a valuable resource. Maintaining genetic diversity is of immense importance for the conservation of ecosystem services

In his best-selling book, *Earth in the Balance,* Al Gore, later Vice President of the United States, wrote in 1992:

> But the single most serious strategic threat to the global food system is the threat of genetic erosion: the loss of germplasm and the increased vulnerability of food crops to their natural enemies. […] it is this supply of genes that is now so endangered [35].

Gene erosion? Gene banks try to record genetic diversity and store it in various ways—not as an electronic datum, but as a biological sample. For the biological sample contains more than just the sequence of nucleotides. It contains the context into which the genetic information belongs. In March 2019, for example, scientists from Russia and Japan showed for the first time that they could elicit biological activity from the nucleus of a **mammoth** that died 28,000 years ago after transfer into a mouse cell. However, it did not become a mammoth [36].

The hereditary material always develops in interaction with the living being that encodes it. Both are nothing without the other. Seen in this light, sequencing the genomes of all living things on Earth certainly helps to gain an understanding of diversity and perhaps to discover interesting or useful new genes. In fact, in the early 2000s, Craig Venter deciphered DNA sequences of microorganisms from the **Sargasso Sea**, an Atlantic marine area east of Florida, and sold the data [37]. Biotechnology companies hoped to discover new catalytic mechanisms in this bag of tricks, which they partially succeeded in doing [38]. But the preservation of genetic diversity is not helped by this. It is more reminiscent of the times when large pharmaceutical companies were still searching for active ingredients for new drugs on a large scale in the rainforest.

The international *Earth BioGenome Project* does indeed aim to sequence the genome of one representative of all higher organisms (eukaryotes) [39]. But even here we cannot speak of a better understanding or even preservation of genetic biodiversity. Let us look at a simple bacterium. The gut bacterium of the species *Escherichia coli* encodes around 2000–4000 genes. In this alone the strains differ. However, the total of all gene variants in all known *Escherichia coli* genomes worldwide, the so-called **pangenome**, amounts to about 18,000 genes [40]. The tragedy is evident in nearly extinct animals, such as the northern **white rhinoceros**. There are only two female animals left, and they seem to be infertile [41]. Now, attempts are being made to preserve the species through artificial insemination. But the gene pool has been narrowed down to two animals. This means that most of the alleles, that is the genetic diversity and thus adaptability, have been lost.

While rhinos are close to extinction due to poaching, natural processes can also contribute to the decimation of a species. For example, a fungus that affects **amphibians** has been blamed for the decline of 500 species over the past 50 years [42]. Of these, 90 species are thought to be extinct, and 124 species have seen their populations shrink by more than 90%. This is the largest documented decline in biodiversity by a single pathogen to date. However, the global amphibian trade is also partly responsible for the fact that the fungus has been able to spread. Not only because we may be facing the greatest extinction of species since the age of the dinosaurs some 65 million years ago (Fig. 7.9), scientists are trying to breed extinct species (Sect. 6.2 and Fig. 6.5) [43, 44].

The fact that genetic material represents a valuable resource of a country and may not simply be exported has been regulated since 2010 in the so-called **Nagoya Protocol**, which expands the aforementioned *Biodiversity Convention*.

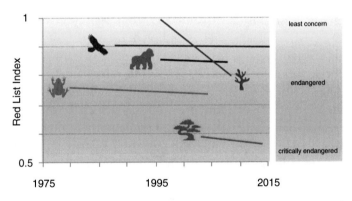

Fig. 7.9 The Red List Index [45] is based on the list of endangered species compiled by the *International Union for Conservation of Nature and Natural Resources* (IUCN). The index measures the change in endangerment over time. Shown is the change for corals, birds, mammals, amphibians and plants. Adapted from WWF [43], data from Butchart et al. [46] and Visconti et al. [47]

Its main purpose is to ensure that when genetic resources, such as certain plant varieties, are used, the countries of origin also benefit and, where appropriate, share in the profits. Expeditions as in colonial times, when natural history museums, botanical gardens, zoos and gene banks were filled to the brim, are thus a thing of the past. On the contrary, there is active negotiation about whether material needs to be returned.

Archiving the entire **genosphere**, that is the totality of all genetic systems that ensure the existence, regeneration and reproduction of the biosphere, is impossible from today's perspective. If only because it is in a state of constant change. But we have already and can continue to deposit small snapshots in gene banks and possibly in the future search for them using completely new methods, such as epigenetic diagnostics (Sect. 8.1).

Increasingly, companies are developing business models for **biobanks** to collect, preserve and distribute living biological material from patients. This can be multiplied in laboratories, stored in cold storage and used to study diseases or test drugs. This methodology is intended to replace animal testing and work on real disease models instead. With the *German Biobank Node* (GBN), the *Federal Ministry of Education and Research* (BMBF) promotes the establishment and networking of biobanks. Biobanks are ultimately also gene banks, since the archived tissues contain genetic information. The value of tissue collected in this way was recently demonstrated when tissue samples from pancreatic cancer patients, one of the most aggressive tumour types, were compared genetically. The result was a **genetic atlas** of pancreatic cancer, which should advance the development of therapeutic methods [48]. The chemist and entrepreneur Jörg Hollidt from Berlin envisions a kind of *Amazon* platform for human cell material. He explains how problematic the project is: Apart from the legal clarification regarding the origin and ownership of the bio-material, there are seemingly trivial issues, such as the very different demands of cellular tissues on culture media and growth conditions. As is often the case, the potential is great, but the implementation is difficult. If we remember (Sect. 5.1) that we are often dealing with a **somatic mosaic** in tissues and, for example, nerve cells even actively rearrange their genetic material, then it may even be impossible and our static view of the genome may proof completely wrong (Sect. 8.3).

References

1. Bersten MC, Folatelli G, García F, et al (2018) A surge of light at the birth of a supernova. Nature 554: 497–499. doi:https://doi.org/10.1038/nature25151
2. Deane-Coe PE, Chu ET, Slavney A, et al (2018) Direct-to-consumer DNA testing of 6,000 dogs reveals 98.6-kb duplication associated with blue eyes and heterochromia in Siberian Huskies. PLoS Genet 14: e1007648. doi:https://doi.org/10.1371/journal.pgen.1007648
3. Kaplanis J, Gordon A, Shor T, et al (2018) Quantitative analysis of population-scale family trees with millions of relatives. Science 360: 171–175. doi:https://doi.org/10.1126/science.aam9309
4. Erlich Y, Shor T, Pe'er I, Carmi S (2018) Identity inference of genomic data using long-range familial searches. Science 362: 690–694. doi:https://doi.org/10.1126/science.aau4832
5. Kaiser J (2018) We will find you: DNA search used to nab Golden State Killer can home in on about 60% of white Americans. In: Science Magazine. Visited 30.03.2019: https://sciencemag.org/news/2018/10/we-will-find-you-dnasearch-used-nab-golden-state-killer-can-home-about-60-white

6. Jiang W, Bikard D, Cox D, et al (2013) RNA-guided editing of bacterial genomes using CRISPR-Cas systems. Nat Biotechnol 31: 233–239. doi:https://doi.org/10.1038/nbt.2508

7. Editorial (2017) Biohackers can boost trust in biology. Nature 552: 291–291. doi:https://doi.org/10.1038/d41586-017-08807-z

8. Huang J, Kang BH, Ishida E, et al (2016) Identification of a CD4-Binding-Site Antibody to HIV that Evolved Near-Pan Neutralization Breadth. Immunity 45: 1108–1121. doi:https://doi.org/10.1016/j.immuni.2016.10.027

9. Lopalco L (2010) CCR5: From Natural Resistance to a New Anti-HIV Strategy. Viruses 2: 574–600. doi:https://doi.org/10.3390/v2020574

10. Smalley E (2018) FDA warns public of dangers of DIY gene therapy. Nat Biotechnol 36: 119–120. doi:https://doi.org/10.1038/nbt0218-119

11. Ebbinghaus, H. (1885). Über das Gedächtnis. Duncker & Humblot, Leipzig

12. Hebb DO (1949) The organization of behavior. Wiley & Sons, New York/USA

13. Tang Y-P, Shimizu E, Dube GR, et al (1999). Genetic enhancement of learning and memory in mice. Nature 401: 63–69. doi:https://doi.org/10.1038/43432

14. Zayner J (2018) BioHack the Planet 2018. In: YouTube. Visited 31.09.2018: youtu.be/2WboOubuI2M und youtu.be/fjGDpEsM13k und youtu.be/CHQleUE-Iwk und youtu.be/ykwR-9MkTZM

15. Clausen R, Longo SB (2012) The Tragedy of the Commodity and the Farce of AquAdvantage Salmon®. Dev Chang 43: 229–251. doi:https://doi.org/10.1111/j.1467-7660.2011.01747.x

16. Canela-Xandri O, Rawlik K, Tenesa A (2018) An atlas of genetic associations in UK Biobank. Nat Genet 50: 1593–1599. doi:https://doi.org/10.1038/s41588-018-0248-z

17. Ye X, Al-Babili S, Klöti A, et al (2000) Engineering the provitamin A (beta-carotene) biosynthetic pathway into

(carotenoid-free) rice endosperm. Science 287: 303–305. doi:https://doi.org/10.1126/science.287.5451.303

18. Zhu Q, Zeng D, Yu S, et al (2018) From Golden Rice to aSTARice: Bioengineering Astaxanthin Biosynthesis in Rice Endosperm. Molecular Plant 11: 1440–1448. doi:https://doi.org/10.1016/j.molp.2018.09.007

19. Vaidyanathan G (2019) Indian court's decision to uphold GM cotton patent could boost industry research. Nature. doi:https://doi.org/10.1038/d41586-019-00177-y

20. Van Dycke L, Van Overwalle G (2017) Genetically Modified Crops and Intellectual Property Law: Interpreting Indian Patents on Bt Cotton in View of the Socio-Political Background. JIPITEC 8: 151–165

21. Gudbjartsson DF, Helgason H, Gudjonsson SA, et al (2015) Large-scale whole-genome sequencing of the Icelandic population. NatGenet47:435–444.doi:https://doi.org/10.1038/ng.3247

22. Geib C (2019) A Chinese province is sequencing 1 million of its residents' genomes. In: NeoScope. Zugegriffen am 14.04.2019:   https://futurism.com/chinese-province-sequencing-1-million-residents-genomes

23. Amann RI, Baichoo S, Blencowe BJ, et al (2019) Toward unrestricted use of public genomic data. Science 363: 350–352. doi:https://doi.org/10.1126/science.aaw1280

24. Carlson B (2010) Medicine could transform healthcare, but payers and physicians are not yet convinced. Biotechnol Healthc 7: 7–8

25. Knowles JW, Ashley EA (2018) Cardiovascular disease: The rise of the genetic risk score. PLOS Med 15: e1002546. doi:https://doi.org/10.1371/journal.pmed.1002546

26. Bahnsen U (2018) Genforschung: Was wird aus mir? Die Zeit, S 33–35

27. Lynch SV, Pedersen O (2016) The Human Intestinal Microbiome in Health and Disease. N Engl J Med 375: 2369–2379. doi:https://doi.org/10.1056/nejmra1600266

28. Franzosa EA, Huang K, Meadow JF, et al (2015) Identifying personal microbiomes using metagenomic codes. Proc Natl

Acad Sci USA 112: E2930-E2938. doi:https://doi.org/10.1073/pnas.1423854112

29. Hegstrand LR, Hine RJ (1986) Variations of brain histamine levels in germ-free and nephrectomized rats. Neurochem Res 11: 185–191. doi:https://doi.org/10.1007/bf00967967

30. Valles-Colomer M, Falony G, Darzi Y, et al (2019) The neuroactive potential of the human gut microbiota in quality of life and depression. Nat Microbiol 13: 1–13. doi:https://doi.org/10.1038/s41564-018-0337-x

31. Wishart DS, Feunang YD, Marcu A, et al (2018) HMDB 4.0: the human metabolome database for 2018. Nucleic Acids Res 46: D608–D617. doi:https://doi.org/10.1093/nar/gkx1089

32. Ma F, Yang Y, Li X, et al (2013) The Association of Sport Performance with ACE and ACTN3 Genetic Polymorphisms: A Systematic Review and Meta-Analysis. PLoS One 8: e54685. doi:https://doi.org/10.1371/journal.pone.0054685

33. Crider KS, Bailey LB, Berry RJ (2011) Folic Acid Food Fortification–Its History, Effect, Concerns, and Future Directions. Nutrients 3: 370–384. doi:https://doi.org/10.3390/nu3030370

34. Lucock M, Yates ZE (2005) Folic acid—vitamin and panacea or genetic time bomb? Nat Rev Genet 6: 235–240. doi:https://doi.org/10.1038/nrg1558

35. Gore Al (2007) Earth in the Balance: Forging a New Common Purpose. Earthscan, New York/USA

36. Yamagata K, Nagai K, Miyamoto H, et al (2019) Signs of biological activities of 28,000-year-old mammoth nuclei in mouse oocytes visualized by live-cell imaging. Sci Rep 9: 4050. doi:https://doi.org/10.1038/41598-019-40546-1

37. Falkowski PG (2004) Shotgun Sequencing in the Sea: A Blast from the Past? Science 304: 58–60. doi:https://doi.org/10.1126/science.1097146

38. Xiang DF, Xu C, Kumaran D, et al (2009) Functional Annotation of Two New Carboxypeptidases from the

Amidohydrolase Superfamily of Enzymes. Biochemistry 48: 4567–4576. doi:https://doi.org/10.1021/bi900453u

39. Lewin HA, Robinson GE, Kress WJ, et al (2018) Earth BioGenome Project: Sequencing life for the future of life. Proc Natl Acad Sci USA 115: 4325–4333. doi:https://doi.org/10.1073/pnas.1720115115

40. Touchon M, Hoede C, Tenaillon O, et al (2009) Organised Genome Dynamics in the *Escherichia coli* Species Results in Highly Diverse Adaptive Paths. PLoS Genet 5: e1000344. doi:https://doi.org/10.1371/journal.pgen.1000344

41. Hildebrandt TB, Hermes R, Colleoni S, et al (2018) Embryos and embryonic stem cells from the white rhinoceros. Nat Commun 9: 2589. doi:https://doi.org/10.1038/41467-018-04959-2

42. Ben C Scheele, Pasmans F, Skerratt LF, et al (2019) Amphibian fungal panzootic causes catastrophic and ongoing loss of biodiversity. Science 363: 1459–1463. doi:https://doi.org/10.1126/science.aav0379

43. WWF (2018) Living Planet Report—2018: Aiming Higher. WWF, Gland, Schweiz

44. Wright DWM (2018) Cloning animals for tourism in the year 2070. Futures 95: 58–75. doi:https://doi.org/10.1016/j.futures.2017.10.002

45. Butchart SHM, Akçakaya HR, Chanson J, et al (2007) Improvements to the Red List Index. PLoS One 2: e140. doi:https://doi.org/10.1371/journal.pone.0000140

46. Butchart SHM, Walpole M, Ben Collen, et al (2010) Global Biodiversity: Indicators of Recent Declines. Science 328: 1164–1168. doi:https://doi.org/10.1126/science.1187512

47. Visconti P, Bakkenes M, Baisero D, et al (2015) Projecting Global Biodiversity Indicators under Future Development Scenarios. Conserv Lett 9: 5–13. doi:https://doi.org/10.1111/conl.12159

48. Raphael BJ, Hruban RH, Aguirre AJ, et al (2017) Integrated Genomic Characterization of Pancreatic Ductal Adenocarcinoma. Cancer Cell 32: 185–203.e13. doi:https://doi.org/10.1016/j.ccell.2017.07.007

Further Reading

Dubock A (2014) The politics of Golden Rice. GM Crops Food 5: 210–222. doi:https://doi.org/10.4161/21645698.2014.967570

Rifkin J (2007) The biotech century: Harnessing the gene and re-making the world. Jeremy P. Tarcher/Putnam, New York/USA.

Smith JS (1990) Patenting the sun. William Morrow & Company, New York/USA.

8
Rethinking Genetics

Several developments and findings in the life sciences demand that we think about genetics in a completely new, or at least different, way. We know better than ever that our environment and our genes are closer to each other than we have suspected. This is what **epigenetics**, which I highlight in Sect. 8.1, teaches us. It is without doubt one of the most exciting areas of genetics and offers, for those who need it, a molecular biological basis for our responsibility for future generations (Fig. 8.1).

Artificial intelligence methods have only been let loose on genetic information on a large scale for a few years. Huge data sets of millions of individuals with the corresponding medical records form part of the basis for analysis. Data from activity trackers like smartphones, smartwatches, and so on provide another part. Without having a goal in mind, the algorithms, distributed on computers all over the world, start calculating. In Sect. 8.2, I give an insight into how far this technology has advanced and what questions the many answers to never-asked questions raise. Ultimately, however, it all boils down to the decision as to whether we want to adapt our genetic make-up in the long-term using genome editing, and according to what criteria.

© Springer-Verlag GmbH Germany, part of Springer Nature 2022   **237**
R. Wünschiers, *Genes, Genomes and Society*,
https://doi.org/10.1007/978-3-662-64081-4_8

Fig. 8.1 It is a truism that not only genes but also environment and culture shape offspring. However, just how profound the molecular mechanisms are has only become clear in recent years. There is a form of inheritance that has an effect far beyond genes: epigenetics

With all the focus on hereditary information and genes, a way of thinking that Dorothy Nelkin and Susan Lindee called *gene essentialism* in their book *The DNA mystique* published in 2004, we easily overlook that the gene is not the only thing [1]. In addition to its epigenetic regulation, the arrangement of genes in the genome also plays an important role. And the genome is subject to **structural dynamics**. In Sect. 8.3, I show that *the* genome does not actually exist. Even within an individual, alterations occur during his lifetime that have nothing to do with the mutations we know so well. These are, of course, also important and a cause of the fact that there will always be antibiotic-resistant pathogens, for example. The engine of evolution drives all human innovation.

8.1 Epigenetics

Epigenetics is without doubt one of the most exciting areas of genetics [2]. The word describes a mechanism of imprinting the genetic material that acts above (Greek *epi*) ordinary genetics. The environment is involved in this in the broadest sense. We can say that the environment acts on the genetics of living beings through mechanisms of epigenetics. This effect—and this is the real kicker—can be transmitted to subsequent generations. And when I write environment in the broadest sense, I do not just mean physical factors like light, temperature, or chemical substances like nutrients or toxins, but also, for example, psychological effects. Let us take a look back at the history of discovery.

In 1742, the young Swedish law student Magnus Ziöberg collected some plants as a hobby on a small island in the archipelago off Stockholm, dried them and archived them as an herbarium. This herbarium found its way to Olof Celcius, professor of botany at Uppsala University and nephew of the inventor of the temperature scale of the same name. He noticed a plant whose leaves, stems and root corresponded to the well-known and well-described **toadflax**, with the Latin name *Linaria vulgaris*. Only the flower puzzled the botanist. Instead of one spur, it had five spurs, and instead of being mirror-symmetrical like a human (with two equal halves, also called zygomorphic or bilaterally symmetrical), it was radially symmetrical like a starfish (Fig. 8.2, left). To make the case, he brought some specimens to his colleague, the most important botanist and naturalist of his time, Carl von Linné. The saying *God created the world, Linné ordered it* shows the latter's important function as the founder of the systematics of living things. At first, Linné thought that a prankster had stuck an alien flower on the toadflax to tease him. When it turned out that

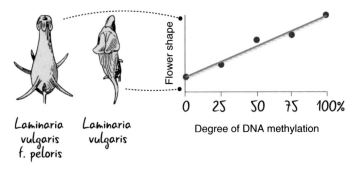

Laninaria   Laninaria
vulgaris   vulgaris
f. peloris

Fig. 8.2 (Left) The radially and bilaterally symmetrical flower forms of the true toadflax *(Linaria vulgaris)* and the monster form *(Linaria vulgaris* forma *peloris)*; after Goethe [4]. (Right) The dependence of flower shape on the degree of methylation (Fig. 8.4) of specific DNA regions. (After Cubas [5])

the plant was intact, he assigned it to the new genus *peloria* (Greek for monster), concluding that the toadflax had produced a monster [3]. He also obtained living plants through Ziöberg, but they quickly died in his botanical garden.

Linné could not solve the riddle of the monster. However, numerous similar examples were found in other plants, which led to the established technical term **pelorism**. The naturalist Charles Darwin already knew about a dozen examples and gradual intermediate stages of flower forms were also found. The botanist Hugo De Vries (Sect. 3.1) was the first of a number of scientists who suspected mutations as the cause of pelorism. It took until 1999 when a completely different cause was found, namely a chemical change, a methylation, at the base cytosine (Fig. 8.2, right). Before we take a closer look at DNA methylation, I should mention the prime example that drew attention to epigenetics in humans.

It is set in Överkalix. This is a small, isolated place in northern Sweden. It became famous in 2001 for a sensational study describing how **diet** can affect heredity. In the

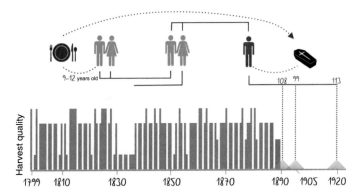

Fig. 8.3 The nutritional status of the grandfather (blue) between the ages of nine and twelve has an effect on the life expectancy of the grandchild. 320 persons born around 1890, 1905 or 1920 were studied. The nutritional status of the ancestors was derived from harvest quality (divided into four levels)

specific **Överkalix study**, medical statisticians examined the influence of diet on the health of first- and second-generation offspring of the inhabitants [6, 7]. The scientists benefited from the availability of crop yields, birth and death certificates, and health data on the subjects (Fig. 8.3). The study examined the living conditions, in particular the nutritional status of the birth cohorts 1890, 1905 and 1920 predicted on the basis of crop yields, as well as the effects on subsequent generations. It was found that malnutrition of male ancestors between the ages of nine and twelve had a positive effect on the life expectancy of second-generation descendants. More specifically, the grandchildren's likelihood of dying from heart disease or diabetes decreases. Wow.

This seems to be **Lamarckism**, that is, the inheritance of traits acquired at lifetime. And that is what it is. It is certainly one of the most remarkable discoveries in the field of genetics in the twentieth century that epigenetics has described a molecular mechanism that supports Lamarck's view of heredity. So, what happens at the level of DNA?

There are several known mechanisms, the most prominent being **DNA methylation** [8]. Since the 1980s, it has been known that certain regions of DNA can be chemically modified by the addition of a so-called methyl group (Fig. 8.4). It has been shown that only cytosines are methylated that are followed by a guanine on the same DNA strand. These are then referred to and written as **CpG dinucleotides** or CpG pairs, to distinguish them from base pairs where the C is on one strand and the G is on the opposite strand, that is a CG base pair. The p stands for phosphate. Methylation has the effect that the structure of the DNA changes somewhat and regulatory proteins and enzymes are unable to bind to the DNA double helix, or bind less effectively. This in turn has far-reaching consequences for the regulation of the genes encoded on the DNA, since the molecular biological apparatus that reads the genetic information (transcription apparatus) is literally denied access. As Fig. 8.4 shows, for example, the enzyme RNA

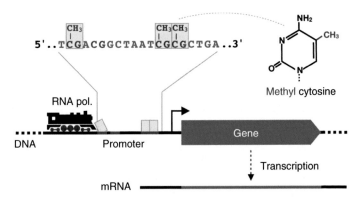

Fig. 8.4 One mechanism of epigenetics is based on the methylation (-CH₃) of cytosines (C) in CpG pairs. These are in the way of the enzyme (RNA polymerase) of transcription, which is why fewer transcripts (mRNA) of the gene are produced and the amount of the gene product is reduced. Methyl groups can be attached and detached depending on external influences

polymerase is prevented from transcribing the gene sequence into a messenger RNA molecule (mRNA) (Fig. 2.4). If the DNA upstream of a gene, the so-called promoter region, is heavily methylated, transcription can even come to a complete standstill: The gene is switched off. Methylated CpG dinucleotides thus cause a change in gene expression and this methylation is dynamic. Methyl groups can therefore be added to and removed from the DNA.

Many plants show epigenetic reactions because they often reproduce vegetatively and can thus easily pass on changes in their DNA to offspring—better: descendants or cuttings. Thus, it is probably no coincidence that the inheritance of acquired traits was first described by botanists such as Jean-Baptiste Lamarck.—Unfortunately, under Stalin there was a massive political misuse of the epigenetic concept by the Russian biologist Trofim Denisovich Lyssenko: It suited the dictator Stalin that one could form a peasant out of anyone by chastisement [9]. But how does information from the environment get into the next generation?

The transmission of acquired characteristics via the germ line first requires an epigenetic modification of the DNA in the sperm and egg cells and its transmission to the embryo [10]. That an embryo is epigenetically imprinted during pregnancy is easy to imagine. However, in the Överkalix case, the information is passed on to the next generation but one. And this is remarkable because until now it was assumed that the germ line is strictly separated from the somatic cells by the so-called **Weismann Barrier** (Fig. 8.5).

German evolutionary biologist and physician August Weismann proposed this strict separation over a hundred years ago. And indeed, it was later found that the epigenetic mark is erased during germ cell formation and after fertilization (known as **germline reprogramming**). However,

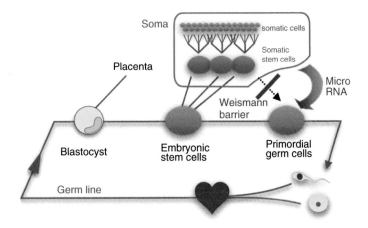

Fig. 8.5 The so-called Weismann Barrier separates the germ line from the somatic cells. According to this, influences on the body *(soma)* cannot affect the primordial germ cells and thus subsequent generations. This hypothesis from the beginning of the nineteenth century has since been disproved

there are exceptions, which we know as maternal or paternal *imprinting* [2]. An underlying mechanism of bypassing reprogramming was elucidated in 2012 [11]. Since then, further mechanisms have been presented as to how epigenetic imprints can act over several generations [12, 13]. A role is also played by short RNA molecules, so-called micro RNAs, which are passed on to the embryo via the sperm and egg cells. These **micro RNAs** are between 17 and 25 nucleotides long and do not code for proteins but regulate the activity of other genes. They can be produced in body cells, transported via the blood and taken up by germ cells. There, the microRNAs can act directly or via DNA methylation, for example. Studies on **endurance athletes** have shown that training has an effect on DNA methylation as well as on the microRNA composition in the blood and—sperm [14, 15]. Surprisingly, genes that play an important role in the nervous system seem to be particularly affected. Indeed, recent results from experiments with mice have shown that physical fitness is transmitted to the offspring via epigenetics and has a positive effect on their **cognitive abilities** [16]. In the study, a group of mice was driven to run regularly in a running wheel within 6 weeks. A control group, on the other hand, remained untrained. The offspring of the athletes showed an altered development of the nervous system during the embryonic phase, had an improved ability to recognize patterns and an increased activity in the power plants of the cells, the mitochondria.

And for the **social sciences**, too, epigenetics suddenly provides a molecular model of how, in the broadest sense, care and welfare can have an effect not only in the present but also in the future. An experiment with rats showed how maternal **breeding behaviour** affects offspring via epigenetic DNA methylation [17]. In the experiment, the mother was constantly present in a control group during the 21-day rearing of the young. In a second group, the mother was separated from her young once a day for 15 min. The result showed that the offspring of the second group were much more active and curious. Molecularly, it was shown that in them the promoter region of a gene was strongly methylated, which codes for a hormone receptor (glucocorticoid receptor) in the hippocampus region of the brain. This receptor is therefore scarcely present in the second group, causing these rats to behave differently, that is to exhibit a different **social phenotype**. What sense can this make? It can be imagined that mother rats that forage outside their nest, encountering enemies, spend more time in the nest rearing. This causes the offspring, when they leave the nest, to be more reticent and not full of curiosity to explore the environment—because they are likely to encounter enemies there. By means of DNA methylation, the mother has given her offspring important information about the environment and corresponding behavioural patterns. My description is, of course, highly simplified. However, it is intended to show that behavioural patterns are the result of an interaction between genetic and environmental factors, which cannot be clearly distinguished from one another.

The potential applications of epigenetics are far-reaching. Scientists at the University of Halle, Germany are currently investigating whether DNA methylation patterns can be used to distinguish conventional from organically grown crops. Epigenetics would thus add another dimension to the genetic fingerprint. This methodology is already being used successfully to detect admixture in saffron [18]. It may even be extendable to confirm the geographical origin of a food sample or to detect impurities. Modern methods have so far been limited to chemical composition, which is often too variable, or DNA sequence, which again are too similar in many cases [19]. I strongly expect that **epigenetic diagnostics** will develop in line with these examples. In a study with rats, for example, it has been shown that contact with certain environmental toxins such as dioxin, pesticides, hydrocarbons such as fuels or plastics leads to specific regions on the chromosomes being methylated [20]. The methylations thus act like a **molecular memory** that can be read out [21].

The interactions between epigenetics and genetic engineering have so far been little researched. However, methods have already been developed, for example to set or delete epigenetic marks using CRISPR/Cas based gene editing [22]. These will play an increasing role in gene therapy for the treatment of diseases such as cancer [23]. In addition, it is hoped that research into epigenetics will provide answers to diseases that currently cannot be explained genetically.

8.2 Artificial Intelligence

Siri, what's the weather going to be like tomorrow? you hear many a person asking these days, and *Chantal, is there a tolerable headache remedy for my genotype?* might be a future question. Siri and Chantal are, of course, smart digital assistants on mobile devices. To answer a question, they must first be understood—in two senses. First, speech must become text. This is done by computer programs that belong to the field of artificial intelligence. However, scientists working in this field do not like this commonly used term—and rightly so. What, pray tell, is intelligence? They prefer to talk about **machine learning**, which is what I would like to do here, too. The computer programs are consequently called **algorithms**, that is a set of instructions for solving a problem [24]. Most algorithms that have been used so far are defined like a mathematical formula: From an input, an output is generated. Period. A corresponding algorithm for speech recognition cannot exactly be described as intelligent, rather as starry-eyed. For example, my surname was happily turned into *Wünsch Dir was* (German for "make a wish") and I had to recite my name to my digital assistant in a way that is completely unnatural: instead of *Wünschiers*, for example, *Wuänschis*. Those days are over, though. Speech recognition has learned, and the basis for this is **learning algorithms**. So, converting speech into text already works very well. The German company *Precire* even offers a service that classifies emotions and the personality of the speaker on the basis of a voice message lasting about a quarter of an hour. Numerous renowned companies, such as Frankfurt Airport or the energy company *RWE,* are already using this service to support recruitment processes [25].

What has been going well for quite a long time is the conversion of writing into text. *Optical character recognition*

(OCR) is also based on learning algorithms. But what else can image recognition do (Fig. 7.2)? Current research projects deal with the detection of diseases from facial expressions [26]. Yes, you read that right. An image processing system called *DeepGestalt* was fed 17,000 images of people who had previously been diagnosed with a disease in the traditional way. In total, the photos were assigned to 200 diseases. The learning algorithm was based on **neural networks**—a method that mimics the function of nerve cells. Recurring patterns lead to signal amplification. After the learning phase, *DeepGestalt* was able to assign diseases to photos with over 90% accuracy—based solely on facial expression or, more accurately, facial shape.

This awakens certain thoughts of medical history in me: at the end of the nineteenth century, the Italian physician and professor of forensic medicine and psychiatry Cesare Lombroso made a significant contribution to the fact that natural scientists were concerned with criminology. Above all, he was convinced that criminals could be recognized by their outward appearance. Thus, he photographed criminals, sorted the pictures by crime and exposed the photos one on top of the other. In this way, photos of the (common) criminal were created, which were then used for identification. For his typification and classification, he used and mixed insights from physiognomics and phrenology as well as ways of thinking of social Darwinism. **Physiognomics** attempts to infer a person's character traits from his or her appearance (physiognomy), especially facial features (Fig. 8.6). In **phrenology**, it is assumed that the externally visible skull shape is shaped in the finest nuances of the shape of the brain, whose shape in turn depends on the activity of the brain areas. This in turn is said to be related to character, knowledge and intelligence. **Social Darwinism** describes a social scientific theory, according to which all

Fig. 8.6 The psychiatrist Guiseppe Antonini described the history of pathognomonics in his book *The Precursors of C. Lombroso* published in 1900. Pathognomonic (nowadays more commonly pathognostic) refers to a symptom that clearly indicates a disease. The description of the symptoms was usually based on animals with characteristics similar to the symptoms. Many a proverb still reminds us of such attributions today. (Source: Antonini [27])

the actions of man are fixed in his biology. Physiognomics and phrenology as methods of connection between appearance (phenotype) and character and the application of knowledge to society and to nurture society bring us very quickly to eugenics. Lombroso's classification of criminals on the basis of external physical characteristics accordingly served the National Socialists, among others, as a template for their racial biological theories—and practices.

Of course (?), it can be assumed that a computer algorithm does not pursue ideological goals. But the possible applications are far-reaching and make the body scanners based on terahertz radiation at security gates seem like children's toys.

Emotions from the voice? Diseases from faces? What about genetic information? Software called *DeepBind* has been optimised to analyse the interaction between DNA or RNA and proteins [28]. This is an important information for identifying proteins that are involved in the regulation of the activity of the genetic material. This in turn can be used to derive molecular diagnostic tests and also drugs. The *ExPecto* program, which uses a DNA sequence to predict in which tissue the coded gene is active in the body and what risk of disease is associated with it, points in a similar direction. Recall that associations between variants in the genome (SNPs) and cognitive traits of individuals have already been found on the basis of association studies (Sect. 4.2 and Fig. 4.10) [29]. Machine learning algorithms go much further. All previous examples were based on what is called **supervised learning**, which means that the algorithm was first trained with expert knowledge. Much more exciting are the algorithms of **unsupervised learning**. These try to recognize patterns independently and thus detach the data analysis from a working hypothesis. Rather, the algorithms lead to new hypotheses. A problem with many algorithms is that it is not possible to trace exactly *how* the correlations were detected. This means that the causality of the correlations must be investigated experimentally afterwards.

Where is the journey heading? More and more data sets will be available for analysis in the future. As described earlier, the *National Health and Medicine Big Data Nanjing Center* of China's Jiangsu Province announced in October 2017 that it would sequence the genomes of one million Chinese people [30]. The *UK Biobank Project* has been collecting data from around 500,000 Britons for over 10 years [31]. In addition to analysing more than 800,000 nucleotide polymorphisms (SNPs) per patient, movement profiles and clinical diagnoses have also been collected. A genetic atlas is now available. As of May 2018, 14,000 deaths, 79,000 cancer cases, and 400,000 hospitalizations had been recorded. The data are constantly updated and are publicly available.

In order to increase the informative value of such analyses in the future, it will be immensely important to collect data on all population strata and ethnic groups. Currently, 78% of the public data sets contain information on Europeans and 10% on Asians [32]. The rest comes from other ethnicities or is unclassified. Data security will also play an increasing role. It is becoming increasingly likely that even anonymised data records can be attributed to individual persons due to the abundance and diversity of the data. In addition to errors in the collection of data, manipulation would also be fatal. For example, scientists from *Ben-Gurion University* in Israel were recently able to show how computer tomography patient data can be manipulated by malware that has been imported into the software system of a hospital [33]. This enabled them to fake cancer diagnoses.

Simply due to the fact that data are often no longer collected on the basis of an indication but as part of large population analyses, many data are quasi surplus. They can no longer be studied at all by scientists in the narrow sense. That's a feast for algorithms: They only need silicon and electricity. And a lot can be expected. In particular, the connection between data of very different kinds—such as DNA sequences, epigenetic status, medical records, behaviour and external life circumstances—will allow increasingly detailed predictions not only in animal and plant breeding, but also in medicine.

It is expected that new insights will be gained into heritability, that is the heritability of traits, characteristics or diseases—but also into the role of environmental influences. Genetic information from the bacteria living on and in us will also play a role. For example, the *Flemish Gut Flora Project* is studying the genomes of bacteria in stool samples from over 1000 subjects [34]. The data show a correlation between subjective quality of life and diagnosed depression on the one hand and the composition of the gut flora on the other (Sect. 7.3). Complex correlations that cannot be explained by means of hypothesis-driven research will be discovered with increasing probability and certainty as a result of self-learning algorithms. Perhaps the digital assistant Chantal will 1 day say:

> Hello, there's a genome update ready for you. Do you want the vector sent to your home address?

The analytical potential of artificial intelligence methods is therefore considerable. But how do we implement this knowledge against the background of the availability of gene editing? In not too long, we may know a manageable, manipulable number of genetic variants in our genome that are associated with highly complex traits such as cognitive and mental characteristics and behaviour. We can diagnose these and use them to select embryos. But we could also use them to interfere with the genome. James Watson, the co-discoverer of DNA structure said in 2002 in an interview on the subject of design babies, *Who wants an ugly baby?* [35] One can counter this: What is ugly? And what are the dangers of a genetic essentialism that sees nothing but genes? [1] I hope we do not (have to) let an artificial intelligence provide us with the answers.

8.3   Dynamic Hereditary Material

I have touched on it before in various places: The genome is not stable. We have to say goodbye to the static picture of a genome that we want to preserve in its current state. Every time a cell divides, changes occur that are passed on to subsequent daughter cells. If this happens in body cells at a late stage of development, such as after birth, it has no effect on subsequent generations. If, on the other hand, the change arises at an early stage of embryo development, then germ cells may also be affected, from which sperm and egg cells arise (Fig. 8.5). Even identical twins have therefore been shown to differ in their genetic information, albeit only minimally (Chap. 2) [36]. In the germ line, on the other hand, this has an immediate effect.

With the still fairly new possibility of analysing the genetic material of a single cell, it was possible to show that some tissues are not genetically identical. This was the assumption, since all cells of a body carry the same genetic material. Instead, it is becoming increasingly clear that we are **somatic mosaics** (see later). In the cells, the genetic information is remodelled [37–39]. This was already known from bacteria and, to a lesser extent, from humans: Genetic remodelling on a small scale forms the basis for the fact that we can form many more different proteins than our genetic makeup would lead us to expect—namely, the antibodies of our immune system. These are proteins that mark so-called epitopes on foreign structures and release them for elimination. About eight amino acids make up their specificity, which, with 20 amino acids, already results in over 25 billion different epitopes. Remember, our genome has 3.2 billion base pairs. Apparently, however, there are much more extensive rearrangements, in which the activity of jumping genes (see later) is also involved [40]. No new information is created, but the existing information is reorganized. We know that this can have great effects from the magnetic puns on the refrigerator wall. These reorganizations seem to play an important role in our nervous system.

It has been shown that the genome of the cells of our nervous system is extremely inhomogeneous [38, 39]. Until now, it was thought that the diversity of cell types and cell-cell connections was mainly due to the different regulation of genes in the genome. However, recent findings show that even the genomes themselves differ. In this context, one speaks of a landscape of *somatic mosaicism*. This means that two or more cells within an individual have different genomes, that is, genotypes. This is known from the egg and sperm cells (Fig. 2.6). This is caused by so-called **mobile elements** in the genome.

Mobile genetic elements can be divided into classes. The most important classes are the so-called Alu-, SVA- (short interspersed element variable number tandem repeat alu) and L1- (long interspersed element 1) elements. They account for about 11% (about 1,100,000 copies), 0.2% (about 2700 copies) and 17% (about 516,000 copies) of our genome, respectively. The differences between the percentages and the number of copies result from the different lengths of the elements of about 300, 3000 and 6000 base pairs. These DNA sequences, also known as **jumping genes**, can change their location in the genome of a cell. They were discovered as early as the late 1940s in maize by the US botanist Barbara McClintock, for which she was awarded the Nobel Prize in 1983. These elements can jump within a chromosome or between chromosomes. If they jump into essential genes, the cell can die. However, since, as we have seen, most of the human genome is plunder DNA, in most cases nothing happens. During jumping, which is also called **retrotransposition**, neighbouring DNA regions can be carried away. This in turn can lead to a rearrangement of genes and a change in their regulation. Mobile elements jump extremely rarely and, as far as is known, mainly at certain stages of development. Depending on when an element jumps, it may occur in more or fewer daughter cells: If the mobile element in a cell jumps during embryonic development, many more successor cells will be generated from the affected cell and a large proportion of the body's cells will carry the corresponding change. Thus, they make up the bulk of the somatic mosaic. If an element does not jump until old age, the change may only affect that one cell or a few descendants. According to current estimates, about 700 neurons are renewed daily in the hippocampus alone. This part of the brain plays an important role in memory. We can therefore speak of a true dynamic of the genome within the lifespan.

And what about in evolutionary time dimensions? A comparison of the proportion of mobile elements in orangutans, chimpanzees and humans shows that the total number of species-specific jumping genes is significantly increased in both orangutans and humans (Fig. 8.7) [40]. However, it is also striking that the proportion of Alu elements decreases from humans to chimpanzees to orangutans. In the same order, **social abilities** also decrease. Combined with the observation that the jumping genes are

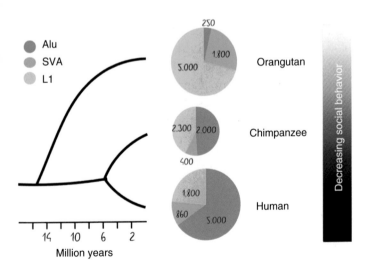

Fig. 8.7 Comparison of the number and distribution of mobile genetic elements specific to apes in each case, and the degree of social behaviour

also active in the nervous system and thus in the brain and contribute to the **neuro-somatic mosaic**, an exciting thesis emerges: Do mobile elements and thus ultimately the dynamics of the genome contribute to social behaviour? This question is currently being intensively investigated in primate research. Thus, a pile of junk DNA would suddenly become a supporting pillar of our humanity.

In Sect. 3.7, we encountered a natural trans-genetic engineering in cockatoo grass, which has incorporated genes from other grass species into its genome via **horizontal gene transfer** [41]. What can we learn from this? Genetic information that we put into the world can be taken up and passed on by organisms. This is a natural process. However, I think we underestimate the regulating powers of nature if we believe that we are destroying ecosystems by crossing out genes. I believe that the spread and release of living organisms by humans, **biological invasion**, is a much more massive intervention, some of which is already having consequences. Invasive species can both increase and decrease the biodiversity of a habitat. In the European Union and other European countries, some 12,000 alien species occur in the environment (plus those plants and animals in parks, zoos and homes), of which an estimated 10–15% are considered invasive. In autumn 2014, the European Union therefore adopted a directive for the first time that regulates both the prevention of the spread and the control of invasive species. There are currently 49 species on the list—such as raccoons, muskrat and nutria, the Asian hornet or the Persian hogweed, the alternate-leaved waterweed or the carrot weed.

I would also like to remind you once again of the development of resistance, for example to herbicides or insect poisons. This is not a weakness of a suitably bred plant or an active ingredient. It is a strength of the dynamic genome to adapt. We can only be successful in dealing with nature if we know about this strength in detail and deal with it. The speed with which biological systems can adapt in the sense of evolution has already been shown in many examples, not only in the case of bacteria. For example, mosquitoes are known to have developed their own subspecies in UK underground tunnels within a few years, which no longer interbreed with mosquitoes living above ground [42, 43].

References

1. Nelkin D, Lindee MS (2004) The DNA mystique: The gene as a cultural icon. University of Michigan Press, Ann Arbor, Michigan/USA. doi:https://doi.org/10.3998/mpub.6769
2. Tucci V, Isles AR, Kelsey G, et al (2019) Genomic Imprinting and Physiological Processes in Mammals. Cell 176: 952–965. doi:https://doi.org/10.1016/j.cell.2019.01.043
3. Gustafsson Å (1979) Linnaeus' Peloria: The history of a monster. Theor Appl Genet 54: 241–248. doi:https://doi.org/10.1007/BF00281206
4. Goethe JW (1820) Nacharbeiten und Sammlungen. In: Troll IW: Goethes Morphologische Schriften Jena
5. Cubas P, Vincent C, Coen E (1999) An epigenetic mutation responsible for natural variation in floral symmetry. Nature 401:157–161. doi:https://doi.org/10.1038/43657
6. Bygren LO, Kaati G, Edvinsson S (2001) Longevity determined by paternal ancestors' nutrition during their slow growth period. Acta Biotheor 49: 53–59
7. Kaati G, Bygren LO, Edvinsson S (2002) Cardiovascular and diabetes mortality determined by nutrition during parents' and grandparents' slow growth period. Eur J Hum Genet 10: 682–688. doi:https://doi.org/10.1038/sj.ejhg.5200859
8. Ehrlich M, Wang R (1981) 5-Methylcytosine in eukaryotic DNA. Science 212: 1350–1357. doi:https://doi.org/10.1126/science.6262918
9. Graham L (2016) Lysenko's Ghost. Harvard University Press, Cambridge, Massachusetts/USA

10. Edith Heard RAM (2014) Transgenerational Epigenetic Inheritance: myths and mechanisms. Cell 157: 95–109. doi:https://doi.org/10.1016/j.cell.2014.02.045

11. Nakamura T, Liu Y-J, Nakashima H, et al (2012) PGC7 binds histone H3K9me2 to protect against conversion of 5mC to 5hmC in early embryos. Nature 486: 415–419. doi:https://doi.org/10.1038/nature11093

12. Eaton SA, Jayasooriah N, Buckland ME, et al (2015) Roll over Weismann: extracellular vesicles in the transgenerational transmission of environmental effects. Epigenomics 7: 1165–1171. doi:https://doi.org/10.2217/epi.15.58

13. Chen Q, Yan W, Duan E (2016) Epigenetic inheritance of acquired traits through sperm RNAs and sperm RNA modifications. Nat Rev Genet 17: 733–743. doi:https://doi.org/10.1038/nrg.2016.106

14. Fernandes J, Arida RM, Gomez-Pinilla F (2017) Physical exercise as an epigenetic modulator of brain plasticity and cognition. Neurosci Biobehav Rev 80: 443–456. doi:https://doi.org/10.1016/j.neubiorev.2017.06.012

15. Ingerslev LR, Donkin I, Fabre O, et al (2018) Endurance training remodels sperm-borne small RNA expression and methylation at neurological gene hotspots. Clin Epigenet 10: 12. doi:https://doi.org/10.1186/s13148-018-0446-7

16. McGreevy KR, Tezanos P, Ferreiro-Villar I, et al (2019) Intergenerational transmission of the positive effects of physical exercise on brain and cognition. Proc Natl Acad Sci USA 3: 201816781. doi:https://doi.org/10.1073/pnas.1816781116

17. Weaver ICG, Cervoni N, Champagne FA, et al (2004) Epigenetic programming by maternal behavior. Nat Neurosci 7: 847–854. doi:https://doi.org/10.1038/nn1276

18. Soffritti G, Busconi M, Sánchez R, et al (2016) Genetic and Epigenetic Approaches for the Possible Detection of Adulteration and Auto-Adulteration in Saffron (*Crocus sativus* L.) Spice. Molecules 21: 343. doi:https://doi.org/10.3390/molecules21030343

19. Hong E, Lee SY, Jeong JY, et al (2017) Modern analytical methods for the detection of food fraud and adulteration by

food category. J Sci Food Agric 97: 3877–3896. doi:https://doi.org/10.1002/jsfa.8364

20. Manikkam M, Guerrero-Bosagna C, Tracey R, et al (2012) Transgenerational Actions of Environmental Compounds on Reproductive Disease and Identification of Epigenetic Biomarkers of Ancestral Exposures. PLoS One 7: e31901. doi:https://doi.org/10.1371/journal.pone.0031901

21. Schmidt F, Cherepkova MY, Platt RJ (2018) Transcriptional recording by CRISPR spacer acquisition from RNA. Nature 562: 380–385. doi:https://doi.org/10.1038/s41586-018-0569-1

22. Pulecio J, Verma N, Mejía-Ramírez E, et al (2017) CRISPR/Cas9-Based Engineering of the Epigenome. Cell Stem Cell 21: 431–447. doi:https://doi.org/10.1016/j.stem.2017.09.006

23. Kelly AD, Issa J-PJ (2017) The promise of epigenetic therapy: reprogramming the cancer epigenome. Curr Opin Genet Dev 42: 68–77. doi:https://doi.org/10.1016/j.gde.2017.03.015

24. Eraslan G, Avsec Ž, Gagneur J, Theis FJ (2019) Deep learning: new computational modelling techniques for genomics. Nat Rev Genet 278: 601. doi:https://doi.org/10.1038/s41576-019-0122-6

25. Rudzio K (2018) Künstliche Intelligenz: Wenn der Roboter die Fragen stellt. Die Zeit 35:22

26. Gurovich Y, Hanani Y, Bar O, et al (2019) Identifying facial phenotypes of genetic disorders using deep learning. Nat Med 25: 60–64. doi:https://doi.org/10.1038/s41591-018-0279-0

27. Antonini G (1900) I precursori di C. Lombroso. Fratelli Bocca Editori, Torino/IT

28. Alipanahi B, Delong A, Weirauch MT, Frey BJ (2015) Predicting the sequence specificities of DNA- and RNA-binding proteins by deep learning. Nat Biotechnol 33: 831–838. doi:https://doi.org/10.1038/nbt.3300

29. Zhou J, Theesfeld CL, Yao K, et al (2018) Deep learning sequence-based *ab initio* prediction of variant effects on expression and disease risk. Nat Genet 50: 1171–1179. doi:https://doi.org/10.1038/s41588-018-0160-6

30. Geib C (2019) A Chinese province is sequencing 1 million of its residents' genomes. In: NeoScope. Visited 14.04.2019: futurism.com/chinese-province-sequencing-1-million-residents-genomes

31. Bycroft C, Freeman C, Petkova D, et al (2018) The UK Biobank resource with deep phenotyping and genomic data. Nature 562: 203–209. doi:https://doi.org/10.1038/s41586-018-0579-z

32. Sirugo G, Williams SM, Tishkoff SA (2019) The Missing Diversity in Human Genetic Studies. Cell 177: 26–31. doi:https://doi.org/10.1016/j.cell.2019.02.048

33. Mirsky Y, Mahler T, Shelef I, Elovici Y (2019) CT-GAN: Malicious Tampering of 3D Medical Imagery using Deep Learning. arxiv.org/abs/1901.03597

34. Valles-Colomer M, Falony G, Darzi Y, et al (2019) The neuroactive potential of the human gut microbiota in quality of life and depression. Nat Microbiol 13: 1–13. doi:https://doi.org/10.1038/s41564-018-0337-x

35. Abraham C (2002) Gene pioneer urges dream of human perfection. In: The Globe and Mail. Visited 18.04.2019: theglobeandmail.com/technology/gene-pioneer-urges-dream-of-human-perfection/article22734105/

36. Weber-Lehmann J, Schilling E, Gradl G, et al (2014) Finding the needle in the haystack: Differentiating "identical" twins in paternity testing and forensics by ultra-deep next generation sequencing. Forensic Sci Int: Genet 9: 42–46. doi:https://doi.org/10.1016/j.fsigen.2013.10.015

37. Fontdevila A (2011) The Dynamic Genome. Oxford University Press, Oxford/UK

38. Carretero-Paulet L, Librado P, Chang T-H, et al (2015) High Gene Family Turnover Rates and Gene Space Adaptation in the Compact Genome of the Carnivorous Plant *Utricularia gibba*. Mol Biol Evol 32: 1284–1295. doi:https://doi.org/10.1093/molbev/msv020

39. Bodea GO, McKelvey EGZ, Faulkner GJ (2018) Retrotransposon-induced mosaicism in the neural genome. Open Biol 8: 180074. doi:https://doi.org/10.1098/rsob.180074

40. Locke DP, Hillier LW, Warren WC, et al (2011) Comparative and demographic analysis of orang-utan genomes. Nature 469: 529–533. doi:https://doi.org/10.1038/nature09687

41. Dunning LT, Olofsson JK, Parisod C, et al (2019) Lateral transfers of large DNA fragments spread functional genes among grasses. Proc Natl Acad Sci USA 116: 4416–4425. doi:https://doi.org/10.1073/pnas.1810031116

42. Byrne K, Nichols RA (1999) Culex pipiens in London Underground tunnels: differentiation between surface and subterranean populations. Heredity 82: 7–15. doi:https://doi.org/10.1038/sj.hdy.6884120

43. Neafsey DE, Waterhouse RM, Abai MR, et al (2015) Highly evolvable malaria vectors: The genomes of 16 *Anopheles* mosquitoes. Science 347: 1258522. doi:https://doi.org/10.1126/science.1258522

Further Reading

Kammerer P (1913) Sind wir Sklaven der Vergangenheit oder Werkmeister der Zukunft? Anzengruber-Verlag Brüder Suschitzky, Wien, Leipzig.

9
Well Then?

We are now at the end of a journey through the world of genes and genomes. Perhaps it was a bit like swimming in the sea: We dived sometimes more and sometimes less deeply into the ocean of knowledge. Whether we ever saw the bottom—certainly not. As a biologist, of course, I find the biological sciences endlessly fascinating. Almost everything I have written about has existed and worked for billions of years. The CRISPR/Cas system is probably similarly old [1]. And yet, to have discovered it and made it applicable as a scientific tool is an achievement. Much like our ancestors created tools—and weapons—from stones billions of years ago. Gene and genome editing will help us to understand the biology and genetics of living things even better. But the digged up knowledge also presents us with new questions. It is an eternal cycle. Answers raise new questions, leading to new answers, and so on. Presumably the cycle is not a cyclone with a calm centre, but rather resembles a galaxy drifting apart. That is what we have to deal with. I imagine you encountered two feelings while reading this: Astonishment and fright.

Astonishment and fright at the diversity of the processes of living things. *Surely none of this can have come about by*

© Springer-Verlag GmbH Germany, part of Springer Nature 2022
R. Wünschiers, *Genes, Genomes and Society*,
https://doi.org/10.1007/978-3-662-64081-4_9

chance, a student once said to me in amazement after a lecture on mechanisms of the regulation of metabolism. Astonishment and fright at what we humans have already researched and to what extent we use the knowledge we have gained to intervene in the genetic material. Astonishment and fright at the amount of information that is currently being processed by self-learning computer algorithms that can learn not only chess and Go, and at the correlations they uncover that we would probably never have thought of ourselves. Astonishment and fright at the idea of what we do not yet know and what is therefore beyond our power of judgement—just think of epigenetics. Astonishment and fright at the fact that it does not seem unreasonable to assume that one day we will be handling gene editing as a citizen science tool in our gardens. And also, astonishment and fright at how close we are once again to the possibilities of use and abuse.

Fright is like the effect of hot wasabi spice: for a moment the sensation is intense, then fading. But what comes afterwards? The good feeling that the fright is over? Shock? Astonishment? Fear? According to Aristotle, wonder becomes the source of inquiry and thought. Fear, however, is usually a bad companion. A healthy amount of fear protects us, but too much inhibits us. The risk of losing can cause fear, while the risk of not winning tends to be viewed neutrally to positively: You could still win, after all. Fear generates rejection. According to Eurobarometer, 90% of Germans and Europeans as a whole have a negative attitude towards genetic engineering—but on a thin ground of knowledge. Today, it is considered politically correct to be against genetic engineering—but that is not enough for the size of the decisions that need to be made.

One question is whether genetic engineering in general and gene editing in particular cause more risks than they can minimise? This question can certainly neither be answered in a general way nor in the same way at all times. We have to assess each individual case and always include a cost-benefit calculation. This also requires research into the risk, that is, the application. And complex problems such as climate change require a holistic approach. That is why we must discuss openly and act together, not in committees or at the regulars' table without publicity, and each as well as he or she can. In line with this, I would like to quote the Swedish physician and data juggler Hans Rosling, who died in 2017:

Let my dataset change your mindset.

Reference

1. Koonin EV, Makarova KS (2019) Origins and evolution of CRISPR-Cas systems. Philos Trans R Soc, B 374: 20180087. doi:https://doi.org/10.1098/rstb.2018.0087

Further Reading

Harari YN (2017) Homo Deus. Vintage, London.
Harrison K (2007) The Fish That Evolved. Metro, London.
Mayr E (1997) This is Biology: The Science of the Living World. Harvard University Press, Cambridge.
Suarez D (2017) Change Agent. Dutton, New York.
Sunstein CR (2005) Laws of Fear: Beyond the Precautionary Principle. Cambridge University Press, New York.

Index

A

Abortion, 132
Adenoviruses, 167
Agriculture, 77
Agrobacterium, 55
AIDS, 80
Algorithms, 247
Allele(s), 14, 19, 113, 119
Amflora potato, 53, 58
Amino acids, 16
Amphibians, 228
Ancestry, 202
Anchoring effect, 5
Animal experiments, 72
Anthropocene, 1
Antibiotic resistance
    genes, 86
Antibodies, 149
Approval procedure, 215
Artemisinin, 188
Artificial intelligence, 237

Ashkenazi Jews, 134
Association studies, 119
Autobiography, 109
Autosomes, 20

B

Bacterium *Rhodococcus*, 53
Banana, 66
Barley, 44
Base pairs (bp), 11, 12
Berlin patient, 151
Big Data, 213
Bioballistic DNA
    transfer, 58
Biobanks, 229
Biodiversity, 202
Biodiversity Convention, 225
Biohackers, 205
Biohack-space, 207
Biological invasion, 256

© Springer-Verlag GmbH Germany, part of Springer Nature 2022
R. Wünschiers, *Genes, Genomes and Society*,
https://doi.org/10.1007/978-3-662-64081-4

Biological safety
    (biosafety), 192
Biosecurity, 193
Bird observatories, 202
Black-or-white fallacy, 5
Blue genetic engineering, 48
Bonnie Orange, 85
Bottom-up approach, 190
Breast cancers, 212
Breeding behaviour, 245
Brown, 49
Bt maize MON810, 53
Bt toxin, 64
Business ethics, 217

C

Cancer cells, 166
Capillary sequencers, 106
Cartagena Protocol, 63, 81
Case-by-case basis, 87
CCR5Δ32 variant, 151
CD4 antigen, 208
Chassis, 189
Chicken egg white, 69
Chinese laws, 157
Christian view, 163
Chromosomes, 12
Cis genetic engineering, 59
Citizen science, 201
Cleaning and cosmetics
    industries, 51
Cloning, 22, 55
Clustered regularly
    interspaced short
    palindromic repeats
    (CRISPR), 143

Cobalt-60 isotope, 42
Codon, 16
Cognitive abilities, 245
Commercial cultivation, 57
Company 454, 108
Completely automated public
    Turing test to tell
    computers and humans
    part (CAPTCHA), 201
Complex circuits, 191
Complexity, 27
Contaminated, 207
Conventional seeds, 216
Cotton, 51
Cousin marriages, 128
CpG dinucleotides, 242
CRISPR-associated genes
    (Cas), 145
Cultivation fees, 216
C-value paradox, 27
Cybernetics, 180
Cystic fibrosis, 119, 133

D

Degree of relatedness, 129
Denisova man, 118
Detergents, 51
Diagnostic examination, 121
Diet, 240
Differentiated discussion, 77
Diploid, 12
Dispute, 143
DIY AIDS therapy, 209
DIY biologists, 205
DIYhplus Wiki, 209
DNA, 9, 54

methylation, 242
modules, 182
polymerase, 105
sequence, 14
synthesis, 177, 193
Dog breeding, 125
Dogs, 202
Dominant, 19, 114
Donor register, 127
Double helix, 11
Duck, 178
DUS testing, 46

E

Easy PGD, 124
Ebola virus, 80, 193
Ecological footprint, 2
Ecosystems, 212
Embryo Protection
    Act, 168
Endurance athletes, 244
Enhancement, 154
Environment, 30
Environmental factors, 83
Enzymes, 15
Epigenetic diagnostics, 246
Epigenetics, 237
Equivalence principle, 61
ES method, 70
European Court of Justice
    (ECJ), 143
European Patent Office, 213
Exceptions, 49
Experimental kit, 206
Explosive TNT, 191
Extreme opponents, 5
Extreme proponents, 5
Ex vivo method, 168

F

Fatty acid composition, 72
Fertility centres, 125
First-generation sequencing
    method, 106
Fitness, 223
Fitness scores, 220
FlavrSavr tomato, 53, 57
Folic acid, 223–224
Forage, 53
Founder population, 116
14-day rule, 169
Functional foods, 223

G

Gaia hypothesis, 4
Gamma gardens, 42
Garage laboratories, 205
Garlic effect, 89
Gene, 18
    drive, 91
    editing, 143
    editors, 145
    erosion, 226
    expression, 16
    gun, 57
    pool, 36, 128
    therapy, 90, 167
Gene Diagnostics Act,
    121, 131
Genentech, 55
Genetics, 18
    analysis providers, 203
    atlas, 229
    data security, 118
    fingerprint, 117
    footprint, 2
    memory, 29

Genetics Diagnostics Act, 221
Genetics Engineering
    Act, 60, 80
Genome editing, 141
Genome synthesis, 185
Genomic scores, 220
Genomics, 20
Genosphere, 229
Genotype(s), 14, 114, 142
Germ cell transplantation
    (GCT), 70, 73
Germ cells, 22, 168
Germline reprogramming, 243
GloFish, 68
Glufosinate (LibertyLink), 64
Glyphosate
    (RoundupReady), 64, 89
Golden Rice, 213
Gonosomes, 20
Grapefruit varieties, 44
Green genetic
    engineering, 48, 52
Grey, 49
GuideRNA, 145
Gut flora, 222

H

Hachimoji-DNA, 181
Haploid, 12
Hazard, 81
Herbicide-resistant plants, 64
Heterosis effect, 46, 216
Heterozygous, 114
Homologous
    recombination, 147
Homozygous, 114

Horizontal gene
    transfer, 86, 256
Human embryos, 150, 169
Hybrid breeding, 46
Hybrid variety, 216
Ice Age, 36
Import, 53
InDels, 148
Indirect risks, 89
Individuality, 170
Information rights, 122
In-market monitoring, 215
Insect-resistant plants, 64
Insemination, 124, 125
Insulin, 55
Intelligence, 135
Intermarry (endogamy), 134
Intracytoplasmic sperm
    injection, 69
In-vitro fertilization (IVF),
    123, 130
In vivo method, 168

J

Juice and wine
    production, 50
Jumping genes, 20, 254

L

Labelling, 63
Labelling requirements, 49
Lamarckism, 241
Large corporations, 215
Law enforcement
    agencies, 204

Laypersons, 91
Learning algorithms, 247
Leigh syndrome, 170
Life expectancy, 203
Ligation, 54
London patient, 151

M

Machine learning, 247
Malaria, 91
Mammoth, 227
Marker-assisted breeding, 46
Markers, 14
Meiosis, 112
Memory, 210
Mendel's Laws of
    Inheritance, 40
Metabolic design, 187
Meta-types, 223
Microbiome, 222
Microclimates, 76
Micro RNAs, 244
Milk, 68
Minimal cells, 189
Misuse (*dual-use
    problem*), 193
Mitochondria, 12, 21
Mitochondrial chromosome
    (mtDNA), 20
Mobile elements, 254
Module, 188
Molecular memory, 246
Monogenic, 119
Monsanto Tribunal, 217
Moratorium, 164
Morphogenetic code, 26

Morphogens, 26
Mosaicism, 149
Mother, 21
Mottled genetic
    engineering, 49
Muslims, 163, 170
Mutagenesis, 41
Mutation, 113
    breeding, 44
    theory, 41
Mycoplasma, 178
Mycoplasma mycoides JCVI-
    syn1.0, 183

N

Nagoya Protocol, 228
Nana and Lulu, 2
Nanopores, 110
Neanderthal man, 109, 118
Neolithic period, 36
Neural networks, 248
Neuro-somatic mosaic, 256
New-born screening, 123
New gene editing, 2
Non-homologous repair, 147
Nucleobases, 11
Nucleotides (nt), 11, 12
Nucleus transfer, 71

O

Off-target effects, 148
One-child policy, 162
Open dialogue, 207
Open-source licensing
    model, 216

Organic rice cultivation, 75
Orthogonal systems, 191
Ötzi, 119
Ötztal Alps, 119
Överkalix study, 241
Overtreatment, 125
Own sperm, 126

P

P4 medicine, 219
Palaeogenetics, 118
Panama disease, 66
Pangenome, 22, 86, 227
Papaya population, 66
Paper industry, 52
Participation, 93
Parts, 187
Patent, 55
Pelorism, 240
Peppermint variety, 43
Performance and
    obedience, 161–162
φX174, 177
Phenotype, 14, 44,
    114, 142
Phrenology, 248
Physiognomics, 248
Plasmids, 54
Point mutations, 61
Polar body diagnostics, 135
Poliovirus, 178
Polygenic, 119
Polymerase chain reaction
    (PCR), 117
Polymorphism, 111, 113
Polyploid, 12
Power of reproductive
    medicine, 129

Precautionary principle, 79
Precision breeding, 46
Predictive examination, 122
Pre-fertility diagnostics, 135
Pre-implantation diagnostics
    (PID), 134
Prenatal diagnostics, 131
Prenatal paternity test, 132
Primitive streak, 169
Prison sentence, 207
Proactive action, 209
Procedure-oriented, 62
Product-oriented, 61
Project Jim, 108
Promoters, 19
Pronuclear injection, 69
Proteins, 15, 111
Protocell research, 190
Protocol, 63
Protoplast fusion, 47
Publications, 162
Pyrosequencing, 108
Pyrrolysine, 180

Q

Quadruplet, 181
Quantified-self-
    movement, 219

R

Race, 116
Radiation, 41
Recall campaign, 86
Recessive, 19, 114
Recombinant DNA
    (rDNA), 54
Recombination, 22

Red genetic engineering, 48
Red Queen Hypothesis, 88
Reduced environmental
    impact, 72
Reforestation, 211
Release, 225
Repair mechanisms, 147
Replication, 112
Research funding, 162
Resistance, 88
Restriction enzymes, 54
Retrotransposition, 254
Retroviruses, 167
Right to know one's own
    parentage, 127
Risk(s), 81, 83
RNA, 16, 110
RNA diagnostics, 122
RNA interference (RNAi), 84
Ruling of the European
    Court of Justice
    (ECJ), 44

S

Safety level 1, 80
Safety level 2, 80
Safety level 3, 80
Safety level 4, 80
Salmon, 68
Sanger sequencing, 105
Sargasso Sea, 227
Saturn, 13
Scientific principle, 79
Second-generation
    sequencing
    technology, 108

Selection, 135
Selection breeding, 36
Selenocysteine, 180
Self-cloning, 80
Sensors, 190
Séralini study, 89
Single nucleotide
    polymorphisms (SNPs),
    111, 113, 117
Size of the genome, 27
SMART breeding, 46
Social abilities, 255
Social Darwinism, 248
Social freezing, 131
Socialization, 159
Social phenotype, 245
Social sciences, 245
Somatic cells, 167
Somatic hybridization, 47
Somatic mosaic(s),
    230, 253
Soybean, 53
Soybean variety, 62
Spanish flu, 193
Special position, 44
Sperm-mediated gene
    transfer, 69
Sports behaviour, 218
Starch, 52
Structural dynamics, 238
Super-Mendelian
    inheritance, 92
Supervised learning, 250
Surrogate sires, 73
Synthetic biology, 181
Synthetic nucleotides, 181
Systems biology, 180

T

Terminators, 19
Territoriality principle, 217
Textile industry, 51
Thalassemia, 132
Third-generation
    sequencing
    machine, 110
Three-parent
    babies, 170
3R principles, 72
Ti plasmid, 56
TILLING method, 47
Toadflax, 239
Tobacco plant, 53, 57
Tobacco variety, 41
Toothpaste, 51
Top-down approach, 189
Transcription, 16
Transcription factors, 25
Transfection, 155
Transformation, 55, 155
Trans genetic engineering, 60
Transgenic petunia
    varieties, 85
Transhumanists, 209
Translation, 16
Transposon mutagenesis, 83
Triplet, 16
Tripronuclear (3PN)
    zygotes, 149
Twins, 112

U

UNESCO Declaration, 160
Unethical behaviour, 126
Unexpected genetic
    changes, 83
Unintentional
    release, 84
Unsupervised learning, 250

V

Value for cultivation and
    use (VCU) testing, 46
Variety approval, 46
Virus resistance, 72

W

Watermarks, 184
Weismann Barrier, 243
Welfare, 162
White genetic engineering, 48
White rhinoceros, 227
World Trade Organization
    (WTO), 217

X

Xenotransplantation,
    67, 186

Z

Zinc-finger nucleases
    (ZFN), 148

Printed in the United States
by Baker & Taylor Publisher Services